The Heritage
of Experimental Embryology

Monographs on the History and Philosophy of Biology

RICHARD BURIAN, RICHARD BURKHARDT, JR.,
RICHARD LEWONTIN, JOHN MAYNARD SMITH

Editors

The Heritage
of Experimental Embryology

Hans Spemann and the Organizer

VIKTOR HAMBURGER

New York   Oxford
OXFORD UNIVERSITY PRESS
1988

Oxford University Press

Oxford   New York   Toronto
Delhi   Bombay   Calcutta   Madras   Karachi
Petaling Jaya   Singapore   Hong Kong   Tokyo
Nairobi   Dar es Salaam   Cape Town
Melbourne   Auckland

and associated companies in
Beirut   Berlin   Ibadan   Nicosia

Library of Congress Cataloging-in-Publication Data
Hamburger, Viktor, 1900–
The heritage of experimental embryology.

(Monographs on the history and philosophy of biology)
Bibliography: p.
Includes index.
1. Embryology, Experimental—History. 2. Organizer
(Embryology)—Research—History. 3. Spemann, Hans,
1869–1941.   I. Title.   II. Series.
QL961.H34   1988   574.3'3'09   87-20416
ISBN 0-19-505110-6
The Appendix, "Hilde Mangold, Co-Discoverer of the Organizer," is reprinted with permission from the
Journal of the History of Biology.

3 5 7 9 8 6 4 2

Printed in the United States of America
on acid-free paper

To Doris and Carola

Preface

Experimental embryology attained a commanding position in the field of Biology during the first half of this century. To biology students of my generation, it held the same fascination as molecular biology and neurobiology do today. We were impressed by the rigorous causal-analytical approach to fundamental problems of embryonic development and intrigued by the prospect of performing experiments on living embryos. We marveled at the elegance and the superb craftsmanship in the performance of the masters of this art and were hardly aware of the incongruity between the enormous complexity of the developmental processes and the limitations imposed by the few techniques at their disposal, which were essentially extirpation, transplantation, and explantation (*in vitro* culture). Spemann's organizer experiment of 1924, which figures prominently in this book, was widely regarded as the crowning achievement of this period.

However, by the middle of the century the resources of experimental embryology were pretty much exhausted. The eclipse of this field is not difficult to understand. For one, the ravages of the Second World War and of Hitlerism took their toll. But the ultimate cause of its decline was rooted deeply in its own axiomatic beliefs and its basic frame of reference. It was built on an organismic, holistic view of embryos and their development. Its aim was to come to an understanding of organogenesis in terms of the potentialities and interactions of embryonic parts which were supracellular entities. It was inevitable that a forceful assertion of reductionist trends would shake its foundations. Indeed the radical shift of emphasis to the cellular and subcellular levels, and, from the 1950's on, to the molecular level, transformed experimental embryology to developmental biology. The brilliant successes of molecular biology drew developmental biology into its orbit. Old-style experimental embryology was doomed; it attained the status of a "classical" science, an epithet which gave it a venerable aura to some, but had a disparaging flavor for others.

If, indeed, experimental embryology has become a chapter in the history of biology, perhaps the time has come to record its history and to take its measure. This book makes a beginning. To write a comprehensive history would have exceeded my resources by far. I have confined myself to that segment of experimental embryology which received its inspiration and direction from the ideas and the experimental acumen of Hans Spemann and which was continued by his associates, his students, his guests from many foreign countries, and by a contingent of distinguished "outsiders" who followed in his footsteps.

A considerable part of the story can be told from my personal experience. I was an eyewitness of—though not an active participant in—the major events that occurred before, during, and after the discovery of the organizer. I was on familiar terms with all the actors in this adventure, and some became close friends. In fact, with the exception of the major figures of the late nineteenth century, A. Weismann and W. Roux and a few others, I was personally acquainted with practically all of the investigators who appear in the pages of this book.

I entered Spemann's Zoological Institute at the University of Freiburg in 1920, simultaneously with Hilde Proescholdt-Mangold, the co-discoverer of the organizer, and Johannes Holtfreter, who was to become Spemann's most original and most productive student—and my lifelong friend. This book owes a great deal to his sharing his insights with me. I spent altogether almost ten years in Freiburg, first as a Ph.D. candidate (I received my Ph.D. in 1925) and from 1928 to 1932 as an *Assistent* (instructor) and *Privatdozent* (untenured faculty member). I spent the two intervening years in the Zoological laboratory of Professor A. Kühn at the University of Göttingen and in Dr. O. Mangold's laboratory of experimental embryology at the Kaiser Wilhelm Institute for Biology in Berlin-Dahlem. When I was Spemann's colleague in my second Freiburg phase, I came to know him well, and the friendship which developed between the mentor and his former student continued by correspondence after my resettlement in the United States in 1932.

I have tried to recapture the atmosphere and the spirit of the Freiburg Institute by occasionally interrupting the scientific discourse and inserting vignettes of my colleagues and friends and others who worked at the Freiburg Institute.

My claim that experimental embryology has vanished into history requires an amendment. While the efforts of the experimental embryologists to subject animal development to causal analysis were remarkably successful, they have left a legacy of unfinished agenda which can be left unattended to for a few decades but should not fall into oblivion. To facilitate the revival of the old issues, when the appropriate time comes, I have attempted to formulate them precisely and to clarify and define the major concepts which are the cornerstones of experimental embryology. I have in mind particularly regulation, embryonic induction, and morphogenetic fields and their self-organization. I have traced the subtle changes to which the concepts were subjected with advancing knowledge, and I have pointed out inconsistencies and pitfalls when the concepts did not fit the data in particularly troublesome situations. In this way I have endeavored to provide a reliable guide that may be useful to developmental biologists who are willing to accept the challenge of redefining the old problems in the new language of cellular and molecular developmental biology and to address them with the sophisticated methodology at their disposal.

Throughout the text, and particularly in matters of theory, I have tried to let the authors speak for themselves. All translations from German publications are mine. The translations from Spemann's book require an explanation. Although all page references in the text of my book are to pages in the English edition (Spemann, 1938), the reader will notice differences in wording. I felt that I should retranslate all quotations from the German edition (Spemann, 1936), because I found the official English text often imprecise and inaccurate.

My friend Dr. Jane Oppenheimer, an experimental embryologist and historian of biology and developmental biology, has read the entire text critically. I owe her a great debt of gratitude for her many comments and improvements of style, and for her encouragement. My friend Dr. Dale Purves has read part of the book. I am grateful for his comments and criticisms. Mrs. Irma Morose has patiently transcribed several drafts of the manuscript to the word processor. Ms. Sue Eads has helped with the bibliography, the legends to the figures and in other ways. Most of the photographs were done by Mr. Joe Hayes of the Department of Anatomy and Neurobiology of Washington University School of Medicine. Some are by Mr. Albert Raccah. Original drawings were done by the Art Department of Washington University School of Medicine. I am grateful to all of them.

Contents

The Heritage
of Experimental Embryology

CHAPTER 1

Introduction

The intriguing phenomenon of animal development has occupied a central position in biology from antiquity to the present. The science of embryology has attracted a succession of great minds, from Aristotle to Hieronymus Fabricius, Marcello Malpighi, William Harvey, Caspar Friedrich Wolff, Carl Ernst von Baer, Wilhelm His, and Wilhelm Roux. Without exception, they were not only superb observers but ingenious interpreters. The subtitle of von Baer's *magnum opus, Ueber Entwicklungsgeschichte der Thiere. Beobachtung und Reflexion* (1828) *(On Developmental History of Animals. Observation and Reflection)*, bespeaks this double commitment. Von Baer's book marks the beginning of modern scientific embryology. The great achievements of descriptive and comparative embryology of the nineteenth century were matched by the pronouncements of major generalizations, such as von Baer's fundamental insight that the embryos of different animal types resemble each other more than do their adult forms. This implies that development proceeds from the sculpturing of general features to the chiseling of fine details which give the species its distinctive characteristics. This "law" served later as one of the cornerstones of Darwin's theory of evolution; it was interpreted as evidence of the origin of different species from common ancestors. Ernst Haeckel's misguided transformation of von Baer's law to his biogenetic law, according to which embryos recapitulate during their development the *adult* forms of their ancestors, set the clock back in more than one way. As Oppenheimer (1967, pp. 156–157) put it:

> Haeckel's greatest disservice, after all, was not his simple ignorance of the morphological exceptions to his law as a descriptive statement, but his emphasis on it as an irrefutable explanation of causal relationships . . . 'Die Phylogenie' he insisted 'ist die mechanische Ursache der Ontogenese' (1891 p. 7) [phylogeny is the mechanical cause of ontogeny], not only distracting to other areas the many

3

who might have otherwise become interested in true mechanical explanation, but refuting, as he thought, irrevocably those who were already involved in developing such interests.

Prominent among the latter was the German anatomist Wilhelm His, who had outlined a program of inquiry into the physiological or mechanical causes of growth and differentiation, starting with simple questions such as the formation of tubular organs from flat epithelia (His, 1874). His voice was silenced by the all-powerful authority of Haeckel—but not for long.

In the 1880s, one of Haeckel's own students, the German anatomist Wilhelm Roux, succeeded in his revolutionary effort to establish the causal-analytical study of animal development as a legitimate branch of embryology. In a way he continued the ideas of His; but his success was based on two original ideas. He realized that the analysis of the proximate causes of development required as a tool the application of the experimental method. He was one of the first to perform experiments on embryos—wisely choosing amphibian embryos as his material (Roux, 1885). He designated the new experimental embryology as "Entwicklungsmechanik" [developmental mechanics]. Some of his followers preferred the name "developmental physiology," however.

Roux's second innovation was theoretical and conceptual in nature. It has to be placed in historical perspective. Since antiquity, two opposing views of the nature of animal development had evolved; they were consolidated in the terms "preformation" and "epigenesis." According to the former, all structures of the organism are preformed in the egg, and development is nothing more than the unfolding and growth of these structures. According to the latter, the egg consists of undifferentiated material, endowed at best with an axial organization; development is a process of the gradual emergence of structures, achieved by manifold external and internal agencies, among them the interactions between parts of the developing embryo. Roux realized that the old antithesis could be recast in causal-analytical concepts. He introduced the two terms "self-differentiation" and "dependent differentiation." Self-differentiation was defined as the capacity of the egg or of any part of the embryo to undergo further differentiation independently of extraneous factors or of neighboring parts in the embryo. Dependent differentiation was defined as being dependent on extraneous stimuli or on other parts of the embryo. These were operational definitions; it was their strength that the alternatives could be tested by experiment, that is, by transplantation and isolation. The two concepts formed a broad frame of reference for analytical experiments in the early days; they were refined and new terms were added (Roux, 1897). The new conceptual dichotomy had an important implication: it removed the adversarial aspect of the old terms, preformation and epigenesis. Now one could ask: To what extent is the differentiation of a given part of the embryo, at a given point in time, self-differentiation, and to what extent is it dependent differentiation?

Roux himself leaned toward the view that self-differentiation was the prevailing mode of development. It was based in part on an experiment that turned out to be flawed (see Fig. 2.2). The pendulum soon swung in the other direction, however. A young German zoologist, Hans Driesch, then aged 24, performed an experiment on sea urchin eggs that revealed entirely unforseen potentials. Every egg cell begins its development by dividing into two cells. Driesch separated the two cells and found to his surprise that each of them formed a complete, though small, larva (Driesch, 1891; see Fig. 2–1). This was the momentous discovery of regulation. Driesch realized the important implication of the result of the experiment: If each half-egg forms a whole rather than a half-embryo, then an interaction must occur between the two cells to restrict their potentials in normal development. Regulation and interactions between parts became the hallmark of development of vertebrates and some invertebrate groups, in addition to sea urchins.

Roux and Driesch excelled as theorists, but their experiments, despite their far-reaching significance, were rather artless. Driesch had shaken the cells apart and Roux had killed one cell of the two-cell stage of the frog with a hot needle but allowed it to remain in contact with the living cell. Further progress depended as much on the refinement of techniques as on the ability to ask the right questions. Two experimental embryologists of Driesch's generation, Hans Spemann in Germany (1869–1941) and Ross G. Harrison in the United States (1870–1959), combined extraordinary experimental skills with superb analytical acumen. They assumed leadership in the new science around the turn of the century and set very high standards of excellence. They both, each in his own way, guided experimental embryology in different directions. Spemann preempted the field of early organ determination. Harrison created the field of experimental neurogenesis and, along with it, the method of tissue culture. He then turned to the analysis of axial determination as related to symmetry and other problems of growth in somewhat older embryos.

This book tells the story of Spemann's ascendancy, culminating in the organizer experiment of 1924. It is also the story of the continuation of Spemann's work by his ingenious student, J. Holtfreter, and his momentous discovery that dead embryonic tissues retain their capacity to induce organized structures. This discovery changed the course of experimental embryology and opened new horizons. I deal with the new directions in the second half of the book. The last chapter is my own synthesis of more recent experiments, presenting the double-gradient theory of the determination of the axial organs. I consider it as the legacy of Spemann's last experiment, the discovery of head and trunk organizer, and thus the attainment of his lifelong goal: to comprehend the embryonic origin of vertebrate organization.

CHAPTER 2

History of
the Organizer Experiment

Experimental embryology was founded by Wilhelm Roux in the 1880s and came of age in the first decades of this century. Its uncontested leaders were Hans Spemann in Europe and Ross Harrison in the United States. The culmination of Spemann's achievements was the organizer experiment, which was published in 1924. Hilde Mangold, who performed the experiment as her Ph.D. thesis,[1] was co-author of the 1924 publication, "Induction of Embryonic Primordia by Implantation of Organizers from a Different Species," which appeared in volume 100 of *Wilhelm Roux' Archiv für Entwicklungsmechanik.*

The experiment occupies a unique place among many important discoveries of the period. The remarkable outcome, combined with the elegant experimental design and the suggestive term coined by Spemann, attracted wide attention.

The experiment was performed on early salamander embryos in the so-called gastrula stage. The gastrula, a hollow sphere composed of several thousand cells, undergoes a complicated process called gastrulation, which, in simplest terms, is the inward movement, or invagination, of the lower (ventral) hemisphere and the simultaneous stretching and spreading of the upper (animal) hemisphere to form a covering for the invaginating parts. Invagination begins at a particular region below the equator. A small groove that marks the site of the incipient invagination is called the blastopore, and the material directly above it is called the upper lip of the blastopore. Spemann and Mangold's experiment consisted of the transplantation of the upper lip from one embryo, the donor, to the flank of another embryo, the host. In order to be able to distinguish between donor and host tissues in microscopic preparations, the unpigmented embryos of one species were chosen as donors and the pigmented embryos of another species as the hosts. To the

---

[1]See the appendix, Hilde Mangold, Co-Discoverer of the Organizer.

surprise of the experimenters, three days after the operation, a nearly complete secondary embryo had formed on the flank of the primary (host) embryo. Microscopic sections showed that the secondary embryo was composed of a mosaic of donor and host cells. The remarkable feat of the upper lip of the blastopore, a small piece of tissue, in producing an integrated whole embryo has earned it the designation of organizer.

At that time, raising an operated embryo outside of its protective jelly membranes was a major achievement; the famous case of *Um 132* (*Um* stands for Urmund [blastopore]), the embryo that lived longest in the experiments of Hilde Mangold, has survived to this day in textbooks of embryology. It had just reached the tail bud stage. Years later, after Johannes Holtfreter had developed a salt solution that permitted much longer survival, and with the benefit of antibiotics, he and others produced by the same experimental design secondary embryos that developed to advanced larval stages and were as complete as their hosts (see Figs. 3–3, 3–4): primary and secondary embryos looked like twins fused at the flank or belly.

This seminal discovery initiated a wide range of further analytical experiments which kept Spemann, his associates, and his Ph.D. candidates busy for a decade. A new chapter began in 1932 and 1933, when J. Holtfreter, a student of Spemann's, working independently in the laboratory of Otto Mangold (Spemann's oldest student and Hilde Mangold's husband), at the Kaiser Wilhelm Institute for Biology in Berlin-Dahlem, made the startling discovery that the upper blastoporal lip retained its inductive capacity after it had been devitalized by heating, freezing, or alcohol treatment. This finding was quickly confirmed by others. Soon thereafter, it was found that a large assortment of animal tissues, both alive and dead, could induce complex structures such as isolated heads, axial organs, and tails. In some instances the induced structures were nearly as well organized as those induced by the upper blastoporal lip. Thus the organizer was relegated from its exalted position to a more lowly one, and the term itself became problematical. Spemann remarked that "a dead 'organizer' is a contradiction in itself" (1938, p. 369).[1] On the positive side, the new data strongly suggested that induction involves the release of chemical agents by the inducing tissues. As a consequence, the emphasis shifted from the inductor to the search for the reacting agent and to the reacting tissue, the gastrula ectoderm, which seemed to have the capacity of responding to chemical stimuli with a wide range of differentiations. Both aspects, the identification of inductive agents and the analysis of the complex responses, have proved to be exceedingly difficult problems. Their pursuit will be taken up later. In the meantime, the dethronement of the organizer has been greeted with sardonic glee by some modern biologists who have tried to deprecate the significance of the organizer experiment and the advances made more recently in the chemical analysis of inductive agents. But does the present preoccupation with cellular and molecular aspects of development and our inability to understand complex supracellular events in this frame of reference justify the depreciation of

---

[1]References which contain only the year of publication but no name refer to publications by Spemann.

the discovery of 1924? I hope I can convince the reader that the issues raised by the organizer experiment are still alive.

Constriction Experiments (1901–1904)

Spemann (1901a, 1902, 1903a) entered the field of experimental embryology at the turn of the century with three publications entitled "Developmental Physiological Studies on the Triton Egg, I–III" (Entwicklungsphysiologische Studien am Triton-Ei, I–III). The experiments were performed on the common European salamander, *Triturus taeniatus.*[1] The title does not indicate that the papers concern constriction experiments on early developmental stages, nor does the introduction to the first paper inform the reader of the intent of the experiments. In fact, it begins with a chapter on "Material and Method," that is, a lengthy description of the relatively simple technique of constricting the egg within its jelly membranes, for which he used a fine baby's hair (from his own son). This attention to minute technical detail is quite characteristic of the author; it is the earliest expression of his perfectionism in craftsmanship which is one of the clues to his later successes.

Spemann was at that time a Privatdozent (untenured faculty member) and assistant in the Department of Zoology of the University of Würzburg in northern Bavaria. His own Swabian origin (he grew up in Stuttgart, the capital of Württemberg) was detectable in the distinct dialect in his speech. The director of the institute was Theodor Boveri, one of the leading cytologists of his time, an eminently creative investigator with a fascinating personality and a strong artistic bent (see the biography by Baltzer, 1962). He became Spemann's friend and mentor. But Boveri cannot be credited with introducing him to experimental embryology. Of the two investigations of Spemann that were instigated by Boveri, one, his Ph.D. thesis, dealt with cell lineage in a parasitic worm, and the other, his Habilitations-Schrift (a requirement for admission to the faculty), was a descriptive study of the development of the middle ear in the frog embryo. Nevertheless, the scientific and personal contact of the two men was close, and as Baltzer, who was at Würzburg during the same years, tells us, Boveri paid a visit to Spemann in his laboratory almost every day.

Spemann tells in his autobiography how he discovered his vocation as an experimental embryologist. A touch of tuberculosis forced him to interrupt his work, and he spent the winter of 1896–97 in Swiss health resorts. The only scientific book he took along was a theoretical treatise by the Freiburg zoologist, August Weismann, one of the most forceful proponents of Darwinism in Germany. Entitled *The Germ Plasm: A Theory of Heredity* (1892), the work was an ingenious attempt to formulate a unified theory of heredity and development, and one of the great theoretical achievements of its time (see E. Mayr, 1982, Chap. 16). A major concern of Weismann's was the problem of how the genetic units in the chromosomes control developmental processes. He considered two alternatives.

---

[1] *Triturus* is the newer designation superseding *Triton,* the name used by Spemann.

The first assumes that the idioplasm (the hereditary material in the nucleus of the egg cell) is split during cleavage by a sequence of unequal nuclear divisions. In this way, the hereditary determinants in the nucleus would be segregated and allocated to different cell groups. They would then determine the specific differentiations in the cell groups in which they resided. Alternatively, the daughter cells of the egg and their descendants might receive the totality of the hereditary material by equal nuclear divisions. In this case, the specific differentiation of a particular tissue or organ would be determined by interactions between nucleus and cytoplasm, or between cells and their neighbors; in this way, specific components of the idioplasm (the chromosomes) would be activated (Weismann, 1892). Weismann opted for the first alternative, which turned out to be erroneous. But the old adage that the error of a genius can be more productive than many uncontested findings of lesser minds is still valid. The lucidity of his discourse suggested specific experimental tests of his hypothesis. Spemann was not the only embryologist to be inspired by Weismann's theoretical speculations and to take up the challenge. He wrote, "I found here a theory of heredity and development elaborated with uncommon perspicacity to its ultimate consequences; but also in the same book some experimental results of the recent past which actually already disproved the theory [of unequal nuclear division]. This stimulated experimental work of my own" (1943, p. 178).[1] Spemann was referring to Hans Driesch's classic experiment on sea urchin eggs, published in 1891, in which Driesch isolated individual cells of the 2- and 4-cell stages by shaking them apart and found that each isolated blastomere could form a complete pluteus larva (Fig. 2–1). This experiment clearly decided in favor of Weismann's second alternative, that is, *equal* distribution of the hereditary determinants. It also revealed an astounding capacity for regulation, that is, the restoration of a whole embryo by reorganization of the half egg. The same result was obtained later for other invertebrate and vertebrate eggs.

The case of the amphibian egg was controversial. The first such experiment was done on the frog's egg by Roux in 1888. It became famous as the "pricking experiment" (Anstichversuch) (Fig. 2–2). Roux had pierced and killed one cell in the 2-cell stage with a hot needle, but did not remove it. He observed that the remaining blastomere formed a half-embryo. (Blastomeres are the individual cells of a cleaving egg.) The result was interpreted as evidence in favor of Weismann's hypothesis of *unequal* nuclear division (Roux, 1888). But Roux's experimental procedure was found to be flawed: when the frog's egg was rotated 180° after killing of one blastomere (Morgan, 1895), or when the two blastomeres were separated completely (Schmidt, 1933), they would regulate and form a whole embryo, as in the sea urchin experiment. In Roux's experiment, the dead blastomere had prevented the living blastomere from undergoing reorganization of the cytoplasm, a prerequisite of regulation. Oskar Hertwig, another leading embryologist of the period, deserves credit for the design of the first experimental isolation of blastomeres and for the fortunate choice of the salamander's egg, which turned out to be a more favorable object for experimental procedures than the frog's egg. Hert-

---

[1] All translations and all brackets [ ] within quotations are by the present author.

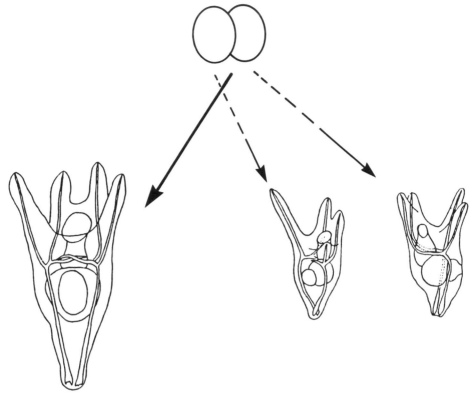

Fig. 2–1. Driesch's experiment of separation of the first two blastomeres of the sea urchin egg. Above, 2-cell stage. Left, normal pluteus larva. Right, each isolated blastomere regulates and forms a half-size pluteus larva. Original drawing.

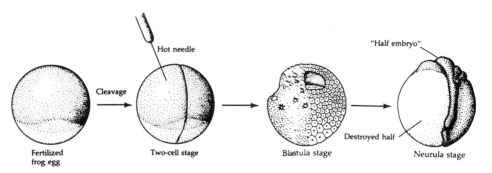

Fig. 2–2. Roux's "pricking experiment." Killing of one blastomere of the frog egg results in a half-embryo. From Gilbert, 1985.

wig tried to separate the two blastomeres by constricting the 2-cell stage in the plane of the cleavage furrow with a fine hair, but he was not successful (O. Hertwig, 1893). Somewhat later, H. Endres (1895) and A. Herlitzka (1897) independently succeeded in obtaining two complete embryos in a few cases, but they did not follow up their findings.

Spemann took over from Hertwig the technique and the material, the salamander embryo, to which he remained faithful throughout his life. In this respect he was not original, nor did he have a clearly defined program. Like others before and after him, he repeated an experiment that intrigued him, guided not by a specific aim but by the intuitive belief that it had the potential of elucidating the problem at hand. His motivation was reinforced by a discovery he had made during the first breeding season. By constricting the egg without separating the blastomeres completely, he had obtained duplications with two heads and one trunk and tail (Fig. 2–3):

> Such animals came to the stage of feeding and it was now most remarkable to see how once the one head and at another time the other caught a small crustacean, how then the food moved through the separate foreguts to the joint posterior intestine. . . . It was probably irrelevant for the well-being of the strange double creature which head had caught the food; it was of benefit to the whole. Nevertheless, one head pushed the other away with its fore legs. Hence two egotisms in the place of one, called forth by the spatial separation of the anlagen. The interest was heightened by the occasional occurrence of such double monsters in man. Here, too, a similar intervention would have the same disquieting consequences. Thus, at last—I was then 28 years old—I had found the beginning of my own scientific journey . . . It was first the fascination of the mystery surrounding the 'partly split individuality,' then the enjoyment of the elegant experimental technique, but then simply the continuing commitment which forced me to seclude myself in my room, one spring after another, and, instead of roaming in the lovely world, to bend over the binocular microscope and tie hairloops around the slippery eggs of salamanders, until I had constricted about a thousand and a half. (1943, pp. 180–181)

It is significant that at the beginning of Spemann's scientific work the experimental results were associated with a psychological phenomenon, and specifically with "individuality." In this instance, individuality was expressed in the behavior and motivation of the two heads in the cases of anterior duplications. But for the embryologist and anatomist, individuality has another connotation: wholeness and integration. The individual organism is the most important organizational unit in the biological world. The embryologist is aware of the continuous identity of the individual, from the fertilized egg to the adult. The most prominent structural manifestations of the individual in vertebrates are the *axial organs:* along the main axis we find brain and spinal cord, skull and vertebral column, flanked in the trunk by segmental muscles and ribs. The axial organs also represent rostrocaudal polarity and bilateral symmetry. In the vertebrate embryo, two transient structures, the notochord (precursor of the vertebral column) and two rows of somites, along with the persisting structure, the neural tube, are the earliest manifestations of axial organs. From Spemann's vantage as an experimental embryologist, the problem of determination of the axial organs became of para-

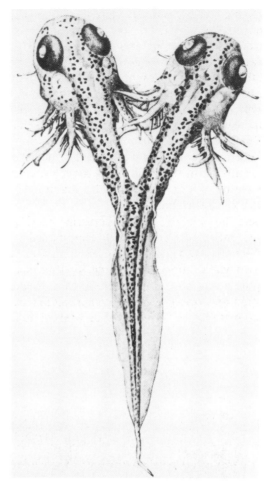

Fig. 2–3. Anterior duplication of salamander larva obtained by partial constriction in 2-cell stage. From Spemann, 1903a.

mount importance; the experimental production of duplications of axial organs became the first step in the analysis of their origin, and the organizer experiment can be considered the last step. Spemann's Inaugural Lecture of 1919 (as the new director of the Zoological Institute of the University of Freiburg) significantly included in its title the term "individuality" ("Experimental investigations of the problem of determination and individuality"). The address ended as follows: "At the beginning of gastrulation, the individuality of the embryo is represented, so-to-speak, by the cells of the upper blastoporal lip which represents the organization center; starting from it, the most important other parts of the body are formed" (1919, p. 591). One notices that this statement and the use of the term "organization center" predate the publication of the organizer experiment by five years. The validity of his statement will be discussed later. At this point I simply want to stress Spemann's commitment to a central issue in analytical embryology that guided him from the beginning.

The first prerequisite for the interpretation of the constriction experiments was to determine the relation of the plane of constriction, i.e., the first cleavage furrow (the constriction at the surface of the dividing egg cell), to the symmetry plane of the future embryo. In the salamander embryo the plane of symmetry cannot be recognized definitively until the blastopore (the site of mesoderm invagination) has made its appearance at the beginning of gastrulation, in the form of a sickle-shaped furrow. The median plane of the embryo bisects the blastopore. Spemann found that in a minority of cases, the first cleavage plane coincided with the median plane of the embryo; in the majority of cases, the constriction separated the dorsal from the ventral part of the embryo. Spemann used the term "frontal" to designate this plane of constriction.

Both the frontal and the median (sagittal) constrictions (Fig. 2–4) contributed to the solution of his main problem, though they guided the analysis in two entirely different directions. By separating the blastomeres in the median plane, Spemann obtained two whole embryos from one egg, thus placing the amphibian embryo in line with the sea urchin and other embryos, and establishing firmly the principles of regulation and equal nuclear division. Different degrees of partial constriction provided material for the in-depth analysis of duplications and other

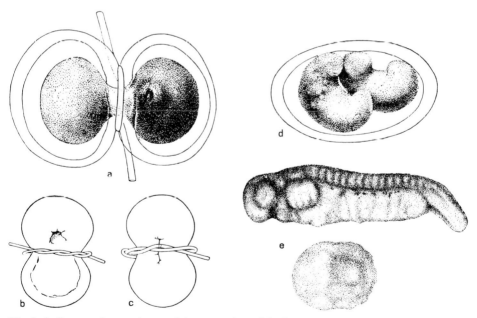

Fig. 2–4. Spemann's experiment of the separation of the first two blastomeres of the salamander egg by constriction. a. The experiment. b,c. Embryos in early gastrula stage with blastopore. b. Frontal constriction separates dorsal from ventral half. c. Median constriction separates left from right half. d. Twin embryos after constriction in median plane, as in (c). e. Embryo and belly piece after frontal constriction, as in (b). From Spemann, 1924a.

anomalies. But in retrospect, the frontal constrictions turned out to have the greater analytical potential. In this instance, only the dorsal blastomere contained the blastopore and underwent gastrulation, eventually giving rise to a complete embryo, whereas the ventral blastomere developed into an ovoid structure, devoid of axial organs. As shown in microscope sections, the derivatives of ventral blastomeres contained three cell sheets which represented the three germ layers. Later on, it was found that this ventral structure, which was called "belly piece" (Bauchstück), occasionally contained kidney tubules and blood islands.

These observations on frontal constrictions evoked the following comment: "If one half of the egg, after termination of gastrulation, is arrested in its further differentiation, although it is perfectly viable, then it follows that with the separation from the other blastomere either the material capable of differentiation [of axial organs], or the impulse [Anstoss] was removed" (1901a, pp. 251–252). And further: "My experiments do not give any information concerning the kind of 'differentiation substance' which is lacking in the ventral blastomere. One can think of an unorganized substance which is necessary either for the release or the formation of [axial] organs; or else it could be organized embryonic material which has the capacity to differentiate into the respective organs and perhaps to incite other cells to differentiate" (1901a, p. 256).

The consideration of all possible implications of an analytical experiment is characteristic of Spemann's discussions; it became much more refined in later publications. But the notion that the formation of axial organs depends critically on a structure, a substance, or a stimulus, located in the dorsal region, was not yet amenable to further analysis, simply because the necessary experimental techniques were not available. The identification of the "dorsal agent" did not come until more than a decade later, when the technique of transplantation on early gastrulae had been developed. But his general theoretical position is clear: differentiation of specific structures depends on interactions with other parts of the embryo.

The general themes of "dependent differentiation," which had been formulated by Roux in the 1880s, and of embryonic induction were in the air. In fact, in theoretical matters, some of Spemann's contemporaries were already further advanced than he. I think particularly of Driesch and his friend Curt Herbst. Following a series of brilliant experiments on sea urchin eggs, Driesch in 1894 (at the age of 27) had elaborated a sophisticated *analytical theory of organic development* in which the role of nucleus and cytoplasm, inductions, chemical stimuli, and other fundamental problems and mechanisms were discussed with great lucidity. It should be noted that Driesch, known to biologists as the proponent of vitalism, was at that time a mechanist. His conversion to vitalism did not occur until 1898 (see Driesch, 1951). Herbst, who deserves credit as the first practitioner of chemical experimental embryology—he made some notable discoveries in his studies of the effects of ions on sea urchin development—had published his theoretical "Programmschrift" (manifesto) on *Formative Stimuli in Animal Ontogeny* in 1901, the same year in which Spemann's first study appeared. It is an extensive and systematic survey of developmental mechanisms. A crucial issue in the approach of both Driesch and Herbst is well formulated by the latter: "to establish the occurrence of formative stimuli which are exerted from one part of the embryo to another, and to demonstrate eventually the possibility of a complete resolution of the entire ontogenesis into a sequence of such inductions" (Herbst, 1901, p. 2). One should not be taken aback by the use of the antiquated term "formative stimulus," which had been taken over from Rudolf Virchow and from botanists of the middle of the last century. Mechanists like Herbst had long divested it of its mystical overtones and used the term synonymously with "inductor." These theoretical considerations had little impact on Spemann, however. He was never given to abstract theoretical speculation. He remarked once that his original ideas for an experiment were often inspired by a concrete visual experience.

I came into personal contact with Driesch and Herbst when I spent two semesters as a student of zoology at the University of Heidelberg in 1919 and 1920. Driesch and Herbst had been close friends since their student days in Jena; they had been among the founders of experimental embryology in the 1890s; they were inseparable at work, at the marine zoological stations in Trieste, Plymouth, Naples, and on many travels to all parts of the globe (Driesch, 1951). They were now in their fifties and distinguished members of the faculty. But in no way did they fit the image of the German Herr Professor; they were informal and unconventional, and both had a good sense of humor. Driesch had long abandoned his experimental work. He had developed his vitalistic philosophy and was now professor of philosophy. I took his course in which his philosophy of the organic was discussed; needless to say, he did not convert me to vitalism. A friend who was his student and wrote a Ph.D. thesis on a topic related to vitalism did not succeed either. He knew Driesch well and conveyed the image of a fascinating, congenial personality; he told me of Driesch's cosmopolitan outlook, his liberal political leanings, and his dedication to pacifism that, later on, brought him on a collision course with the Nazi regime. Ultimately Driesch lost his professorship at the University of Leipzig. In his later years he became deeply involved in parapsychology and occultism. I remember him as a spirited lecturer, but his drawings of planarians and other creatures on the blackboard showed clearly that his talents were in abstract thinking and not, like Spemann's, in the artistic presentation of the visually concrete.

I spent most of my time in the zoological laboratory, which was under the directorship of Curt Herbst. I have mentioned that he had an excellent reputation as an experimental embryologist. As a person he was amiable and easygoing, but difficult to approach. My aunt, Dr. Clara Hamburger, one of the first women Ph.D.s in zoology in Germany, was a close associate of Herbst's predecessor, Otto Bütschli, and a faculty member in the institute. She brought me into personal contact with Herbst. In spite of this, I never got to know him well. He was aloof and a confirmed bachelor who shared his social life with only a few trusted friends. I learned a great deal of zoology in his lectures and laboratory courses, and he had a lasting influence on my life and career through one of his seminars. Through the good offices of my aunt and a newly acquired friend, Walter Landauer, who worked on his Ph.D. thesis under Herbst, I, a second-year student, was admitted to a graduate seminar in experimental embryology which was attended by fewer than a dozen advanced students. We presented papers on current research in this field, followed by lively discussions. It was this seminar that decided my fate—I was to become an experimental embryologist. I was then not bothered by the fact that while I was conversant with the concepts and problems of analytical embryology, I had only the most perfunctory knowledge of actual embryos. I made up for this deficiency in the following year in Freiburg. The flexibility of the curriculum which prevailed then at German universities did have its advantages! Incidentally, Walter Landauer, who became disillusioned by the political atmosphere in Germany, emigrated to the United States soon after he had earned his Ph.D. and became a leader in developmental genetics and teratology. Our common interest in chick embryos brought us together again after my arrival in the United States, and in the 1940s he provided the material for my experiments on the creeper mutant.

The analytical potential of the constriction method was limited. The method permitted only variations in the degree of constriction, the plane of constriction, and the stage of development. Spemann exploited these possibilities to the fullest, and, as I have pointed out, the frontal constrictions raised questions which laid the foundation of later work. Of other results I mention only a few highlights. Considerable insight was gained into the process of amphibian gastrulation, which was then poorly understood. (Gastrulation is the process by which the lower hemisphere of the hollow embryo, called the blastula, is invaginated and the upper hemisphere stretches and flattens to cover the invaginated parts.) The important role of invagination, which was decidedly underestimated by many, was firmly established. Spemann also made substantial contributions to teratology, which was at that time held in much higher esteem that it is now. Embryologists considered it one of their tasks to explain bizarre monsters which occurred spontaneously in domestic animals and man and attracted wide attention. By varying the degree of constriction, Spemann succeeded in producing a graded series of anterior duplications. A medium-deep constriction in the median plane gives two complete heads; if the constriction is somewhat deeper, there may also be two pairs of fore limbs and a merging of the two parts in one posterior trunk and tail with a single pair of hind limbs. If the constriction is very slight, then only the anterior head is duplicated; and cases are encountered in which the two inner eyes are fused, or even monsters with three eyes. On the other hand, if the constriction cuts very deeply, and only a narrow bridge connects the two blastomeres, then the two embryos are fused only at the region of the anus (see Spemann, 1903a). And, as was mentioned, a complete separation of the two blastomeres results in identical twins. All this seems straightforward, but why do the partial constrictions give rise only to *anterior* duplications? Spemann's insight into the gastrulation process enabled him to answer this question. It has to be realized that the blastopore, where invagination begins, represents the posterior end of the future embryo. Hence the material that invaginates first will become head mesoderm, which will induce the brain and eyes. While gastrulation proceeds and while this material moves forward, it is cleaved by the constriction furrow "like the river which is divided by the pier of a bridge" (1903a, p. 598). The deeper the furrow, the more the mesoderm mantle is divided.

Varying the stage of development at which the constrictions were performed led to a conclusion of great importance. Whereas constrictions in the early gastrula still produced anterior duplications, the capacity for regulation decreased with the progression of gastrulation, and constrictions in the early neural plate stage resulted merely in an indentation without duplication. Since regulation implies that embryonic parts can give rise to structures different from those which they would form in normal development, the loss of regulation capacity at the end of gastrulation means that the axial organs become irreversibly determined during gastrulation (1903a). The constriction experiments came to an end with an investigation of the origin of cyclopean (single) eyes following very slight constrictions (1904). The method was used once more in a Ph.D. thesis of Spemann's student Hermann Falkenberg. He analyzed the occurrence of *situs inversus* (mirror imaging of normal asymmetry) in heart and liver of identical twins and anterior duplications following constriction. Falkenberg lost his life in the First World War, and Spemann wrote up the results (Spemann and Falkenberg, 1919).

Where did Spemann stand in 1904 at the end of the first period of his scientific adventure? The problems of gastrulation and of the origin of duplications had been greatly advanced, although many special problems remained unresolved, as for instance, the question of whether the two eyes and lenses originate from two separate anlagen or by fission of a single anlage. Concerning his major theme, the origin of the axial organs, he had obtained two important pieces of information: (1) an essential factor for the realization of axial organs is localized in the dorsal region; and (2) the determination of axial organs occurs during gastrulation. But further progress was stymied by the limitations of the constriction method; its potential had been exhausted. It was necessary to take the embryos out of their jelly membranes and to develop a technique by which the gastrulae could be manipulated experimentally. The invention of the glass needle technique solved this problem.

I have mentioned that Spemann's early interest in anterior duplications was heightened when he observed that the two brotherly heads competed for prey. He apparently lost sight of this aspect completely. Obviously, the many analytical problems which he uncovered provided an abiding challenge and greater excitement than the bizarre behavior.

Lens Induction (1901–1912)

In the 1880s, Roux had formulated the major goal of experimental embryology: to explore the causation of differentiation in different parts of the developing embryo. He had realized that interactions between parts play a major role and that the analytical experiment can reveal such causal relations. But sooner or later, any given part would become independent of extraneous agents and capable of "self-differentiation," that is, typical differentiation according to its normal fate. To give the term a precise meaning, the stage of development at which self-differentiation is attained has to be strictly defined. Roux's terms "self-differentiation" and "dependent differentiation" are not alternatives, as is surmised occasionally, but complementary.

A specific path of differentiation can be initiated in an embryonic part by a stimulus from an adjacent region. The term "induction" had been used by Driesch and Herbst to denote such an interaction, but the concept was rather vague and rigorous experimental evidence was lacking. Spemann addressed this general problem in a series of experiments on the eye and lens which were begun shortly after he had started the constriction project. His intent is stated in the introduction to the first publication on this topic:

> The complex apparatus of the vertebrate eye originates by a sequence of developmental processes which . . . are interlocked spatially and temporally. The following is an experimental contribution to the question of whether these developmental processes proceed dependently or independently of each other; that is, whether their spatial and temporal coordination is guaranteed by a causal interaction or by a harmony which dates back to earlier stages, and perhaps to the egg. (1901b, pp. 61–62)

Originally he had planned to analyze the transformation of the optic vesicle into the optic cup, lens formation, and the factors responsible for the transparency of the cornea. Eventually, his inclination to go into depth rather than breadth prevailed, and he focused on lens formation. This proved to be a fortunate choice: he opened up a new field, and, to this day, lens induction is a paradigm for induction in general.

Of course, the normal development of eye and lens was well known in 1901. Following the transformation of the neural plate into a tube by upfolding and fusion of its marginal folds, two evaginations of the anterior end of the neural tube form the optic vesicles. The lens forms at the point where they make contact with the overlying epidermis. The optic vesicle is transformed into a cup by an infolding of its outer half; the epidermal cells destined for lens differentiation then form a vesicle which is pinched off from the epidermis and fits neatly in the optic cup (Fig. 2–5). It took an analytical mind to wonder whether the optic vesicle plays the role of a causal agent in lens formation, and visual imagination to design an experiment that would solve the problem.

In the neural plate stage (Fig. 2–8), the prospective eye anlagen are situated near the midline of the anterior plate, whereas the prospective lens-forming cells are located in the lateral epidermis, at some distance from the margin of the plate. Hence the eye-forming area can be removed without impairing the lens-forming cells. Spemann performed this experiment on the embryo of the European frog, *Rana fusca,* using the rather crude technique of destroying the prospective eye anlage with a hot needle, or electrocautery. Microscopic inspection a few days later showed that both eye and lens were missing on the operated side (Fig. 2–5). The obvious conclusion was that a stimulus from the optic vesicle is required for lens formation. This conclusion was supported by the observation that in cases in which the eye had not been destroyed completely but merely reduced in size, lenses had formed when the eye rudiment had come into contact with the epidermis but were absent when the eye rudiment was too small to reach the epidermis (Spemann, 1901b). However, this experiment did not answer the question of whether the optic vesicle merely served to trigger lens formation or played an instructive role. To decide this issue Spemann suggested two experiments: to bring the optic vesicle in contact with foreign epidermis by transplanting it under

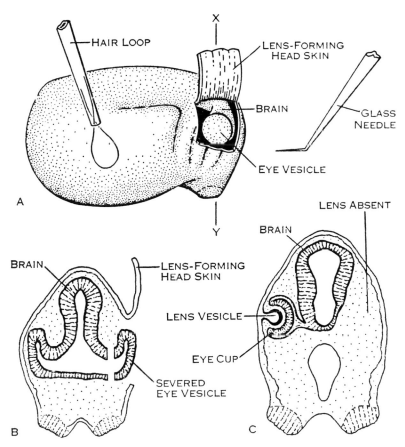

Fig. 2–5. Extirpation of right optic vesicle. A. Skin over optic vesicle folded back, optic vesicle exposed. B. Cross section in X—Y plane of A. Optic vesicle severed with glass needle. C. Later stage. Normal optic cup and invaginating lens vesicle on left side; both missing on right side. From Hamburger, 1960.

flank epidermis or by replacing the epidermis over the optic vesicle with flank epidermis.

The American embryologist Warren Lewis had independently conceived the same idea and actually performed both experiments on the American frog, *Rana palustris* (Lewis, 1904). In both instances, typical lenses differentiated from flank epidermis. The experiment thus decided in favor of an instructive rather than a mere triggering role on the part of the optic vesicle. Spemann and Lewis deserve credit for the first experimental documentation of a case of embryonic induction, which turned out later to be an important and widespread mechanism in epigenetic development. These experiments were also a milestone on the road to the organizer experiment, since inductions are an integral part of the formation of the secondary embryos produced by the organizer.

Strangely enough, neither Spemann nor Lewis used the term induction in reporting his results, in spite of the fact that it had been introduced in experimental embryology much earlier. As was mentioned, Driesch used the term in his theoretical treatise (1894), though in a rather vague and ill-defined way, as a collective term for a number of unrelated causal relations (see Oppenheimer, 1970). Perhaps the lack of a precise definition discouraged its use by Spemann and Lewis; it fully entered the vocabulary of experimental embryology much later, in the 1920s, when it was clearly understood to mean "an embryonic activity which determined the cytological fate of the reacting cells" (J. Holtfreter and Hamburger, 1955, p. 275).

Soon it turned out that the original notion of lens induction was a considerable oversimplification. Mencl (1903) found a double-headed salmon embryo in which two lenses had differentiated in the absence of optic cups. At first Spemann tried to brush aside this observation with a somewhat contrived interpretation (1903b). But when he repeated his original experiment on the green frog, *Rana esculenta,* a close relative of the species he had used earlier, the brown frog, he found to his great surprise that here also well-differentiated lenses had formed after successful extirpation of the optic anlage in the neural plate. It seemed hardly credible that different causal mechanisms should be involved in the development of a structure that is common to all vertebrates. In subsequent years, however, other instances of self-differentiation of lenses were discovered, and one had to accommodate the improbable. To complicate matters further, the capacity of foreign epidermis to respond to the inductive stimulus was not universal, either. In some instances, only head epidermis adjacent to the eye was competent to do so, and in others, lens-forming competence was restricted entirely to the epidermis covering the optic vesicle. Moreover, the simplistic notion of a one cause–one effect mechanism, which has occasionally misled not only the early experimental embryologists, was soon dispelled. As will be shown, head mesoderm and other cofactors were also implicated in lens differentiation. In one instance, even the temperature at which the embryos were raised before the actual eye extirpation was performed was found to influence the outcome (Ten Cate, 1956).

Eventually, a consensus was reached that there is a wide range from complete independence of lens differentiation to complete dependence on the optic vesicle. In contrast, the capacity of the optic vesicle to induce a lens seems to have been retained even in those instances in which the lens is capable of self-differentiation. Filatow (1925) removed the lens epithelium of a *Rana esculenta* embryo, which is capable of self-differentiation, and replaced it with flank epidermis of a toad, which can respond to the lens-inducing stimulus; in this circumstance, a normal lens was formed. This meant that the optic vesicle has retained its lens-inducing capacity even though its own epithelium did not "need" it. I mention this seemingly minor point because it has a bearing on a particular aspect of the organizer experiment.

It became clear early in the course of the analysis of the lens problem that further progress depended on improvements of the available techniques. In particular, when transplantation experiments became indispensable they required special delicate instruments. W. Lewis had used a pair of fine scissors and sharpened steel needles. Spemann invented a glass needle technique for microsurgery which he described in 1906. For him, this was a breakthrough of signal importance; apart from its application to the lens problem, the later transplantation experiments on gastrulae which culminated in the organizer experiment would not have been feasible without this technique. It proved to be very versatile; it soon came into general use and was adopted for other forms, such as chick embryos.

We learned from Spemann's assistants how to prepare our own tools. The glass needles (Fig. 2–5) were fashioned by holding a glass rod over a Bunsen burner until the heated middle part was softened. It was pulled out to a very thin thread which was then broken off. The thin glass fiber was drawn out further to an almost microscopic diameter over a microburner. The very fine tapering end piece of the glass needle was attached to a handle, a tapering glass rod, at an angle of 120°. The microburner was another of Spemann's inventions. It was prepared from a finely drawn-out piece of glass tubing and connected to a gas outlet by a piece of rubber tubing provided with a clamp to control the flow of gas. The glass needles were part of an arsenal of auxiliary tools, also designed by Spemann; hair loops (Fig. 2–5) for the manipulation of embryos were made of fine baby's hair, sealed with wax into the tip of a finely drawn-out glass tube; its opening was just large enough to accommodate the two ends of the hair. A capillary micropipette was used to transfer the tiny transplants. It had a lateral hole at its broad extension which was covered with a rubber membrane; pressure on the latter with the thumb permitted minimal suction. Bridges made of coverslips served to hold transplants in place until they had healed in. These tools were very inexpensive; so were the glass dishes in which we reared the embryos. We spent the equivalent of only a few dollars during the entire breeding season. Our most expensive instruments were the fine watchmaker's forceps imported from Switzerland. They were used for many purposes, from removing the fine vitelline membrane around the egg—a tricky business—to threading the hair loops (see Hamburger, 1960). Manufacturing the instruments required considerable skill, and as students we spent a good deal of time perfecting it. Spemann, who was not particularly strict in other matters, became very critical when we did not live up to his standard of perfection in this exercise.

Spemann took special delight and great pride in the invention of these gadgets. "From my father's side came an inclination and talent for everything that had to do with craftsmanship. . . . What my profession required in this respect was therefore not a burden but a pleasure which let me forget time, and I had to guard myself, that it did not grow beyond control and become an end in itself. I found particular satisfaction in the invention of small tools and gadgets for my experiments which, if possible, should be simple enough to be prepared without outside help" (1943, p. 202). No doubt, the deliberate cultivation of his native talent for manual craftsmanship and his perfectionism contributed substantially to his successes in experimental embryology. Yet others achieved the same end without much enthusiasm for techniques. For instance, I was told by a close collaborator of R. G. Harrison that he was not at all interested in the technical aspects of his work. Yet Harrison did wonders with his favorite tool, a pair of iridectomy scissors which he had taken over from the workshop of ophthalmologists. Others did well with sharpened steel or tungsten needles.

There was one weak spot in Spemann's resources: his technical acumen was not matched by a proficiency in chemistry. What was wanting was a friendly colleague in the chemistry department—across the street from zoology in Freiburg—or, for that matter, in his own department, who could have advised him on the simple matter of an adequate culture medium for his embryos. This could have saved the lives of hundreds of operated embryos and many more hours of labor. As it was, we used filtered tap water! In his years in Berlin and, as I remember, in the early days in Freiburg, Spemann had water shipped in from Würzburg. The magic ingredient of the "Würzburger Wasser" was apparently its high calcium content. It was not until 1931 that Holtfreter found the balanced salt solution which solved the problem and saved the embryos, though it came a bit too late for Spemann.

Determination at the Gastrula Stage (1914–1919)

Spemann spent altogether fourteen years at his alma mater, the University of Würzburg, in the intellectual sphere of Theodor Boveri. His natural propensity to focus on basic problems, yet to attend meticulously to detail, was undoubtedly reinforced by Boveri, with whom he shared an interest in problems of heredity and development. Their actual research programs did not intersect, however. Boveri was essentially a cell biologist whose major concern was the role of the nucleus and chromosomes in development. Spemann, as we have seen, adopted the causal-analytical approach of Roux, and in his analysis of embryonic induction and related phenomena was faced with the interactions of supracellular units, although, of course, he could not always ignore cellular events.

Spemann obtained his first independent position in 1908, when he was called to the small University of Rostock, on the Baltic Sea, as the director of its Zoological Institute. During the six years he spent there, his work on lens induction was completed with a lengthy publication (1912a). A project on the polarity of the inner ear vesicle (1910) and another one on *situs inversus* (mirror image of asymmetrical organs, such as heart and liver) (1912b) had little impact. During this period no new ground was broken.

All this changed with his appointment, in 1914, as the co-director and head of the Division of Developmental Mechanics of the Kaiser Wilhelm Institute for Biology in Dahlem, a suburb of Berlin. The new environment was exceptionally conducive to research. The dozen or so Kaiser Wilhelm Institutes were at that time the only institutes in Germany devoted exclusively to pure and applied research. Spemann found a number of distinguished scientists among his colleagues: Carl Correns, the co-director of the institute, a leading plant geneticist and one of the rediscoverers of Mendel's laws; the animal geneticist Richard Goldschmidt; the protozoologist Max Hartmann; and the biochemist Otto Warburg. Within walking distance were other institutes, for physics, chemistry, etc., inhabited by a world-renowned elite. The four and a half years he spent there were among the most productive of his life, during which he moved from relative obscurity to wide recognition.

Spemann moved into his own house in Dahlem in September 1914. The construction of the laboratory was delayed by the outbreak of the war, and Spemann volunteered for several months at a military hospital in Dahlem, until the institute was ready for occupancy early in 1915. He was too old for military service and was able to pursue his investigations relatively undisturbed throughout the war. However, his health deteriorated toward the end of the war. He suffered from undernourishment, since he refused to procure food on the black market (almost everybody else did it), and after the war it took long periods at a south German spa for him to recuperate.

Before Spemann left Dahlem in the spring of 1919, one of his major contributions appeared in print under the title "On the determination of the first organ anlagen of the amphibian embryo" (1918). With this work he returned to the central topic that had emerged from the constriction experiments: the determination of the axial organs.

For an appreciation of the new experiments, a clear understanding of the early phases of amphibian development, gastrulation, and neurulation, and of the structure of the tail bud stage is indispensable. The pregastrulation embryo, called the blastula, is a hollow sphere; its cavity is the blastocoele and a mass of yolk-laden cells forms its floor. Gastrulation is the process of invagination of the ventral hemisphere into the interior, while the dorsal hemisphere stretches and flattens and forms the covering of the invaginated parts. (Embryos in successive stages of gastrulation are pictured in Fig. 2–6.) Invagination begins at a particular region below the equator; the site of invagination is marked by a small slit, the blastopore. Adjacent cells begin to invaginate and the blastopore becomes sickle-shaped. At that stage, bilateral symmetry and the axial organization become vis-

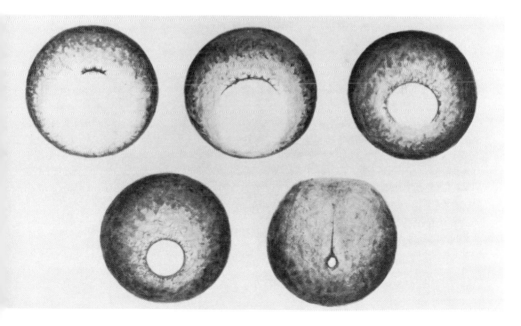

Fig. 2–6. Gastrulation in frog embryo. The incipient blastopore (upper left) becomes sickle-shaped, horseshoe-shaped, and circular and closes eventually. From Balinsky, 1978.

ible. The plane that bisects the blastopore is the median plane of the embryo. The region above the blastopore is designated as the dorsal lip; thus dorsal and ventral are identified. The blastopore marks the posterior end of the embryo, and the invaginating material moves forward toward the anterior end. The site of invagination extends laterally and the blastopore becomes horseshoe-shaped; eventually it becomes circular. It encloses the yolk-laden cells, which are called the yolk plug. The latter becomes smaller, and gastrulation ends when all material has disappeared in the interior and the blastopore is reduced to a small slit, which is the site of the future anus. The axial organs will develop along a meridian which is marked by the blastopore, the upper (animal) pole, and the head region on the side opposite the blastopore.

The changes that go on in the invaginating material are very complex; they can be dealt with only in broad outline (Figs. 2–7, 2–8). While invagination proceeds, the invaginating material splits. The ventral part, composed of yolk-laden cells, forms a trough called the archenteron because it is the precursor of the intestine. The lateral walls of the trough grow upward and eventually fuse in the dorsal midline; in this way the intestine is formed. At a stage when the trough is still open, a sheet of cells has split off from the yolk-laden cells; it spreads forward and sideways and forms a temporary roof over the trough which has earned it the designation of archenteron roof. The forward movement of the archenteron results in the gradual elimination of the blastocoele.

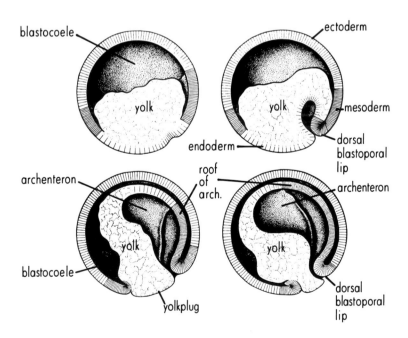

Fig. 2–7. Schema of gastrulation in amphibians, seen from medial plane. Note invagination of mesoderm and formation of archenteron (primitive gut), replacement of blastocoele by archenteron. From Spemann, 1936, after drawings by V. Hamburger and B. Mayer.

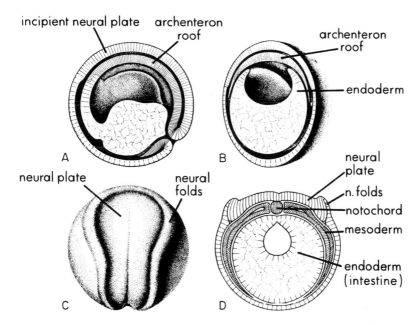

incipient neural plate archenteron
roof

archenteron
roof

endoderm

A

B

neural plate

neural plate

neural
folds

n. folds
notochord
mesoderm

endoderm
(intestine)

C

D

Fig. 2–8. Schema of beginning neurulation in amphibians. A. View from median plane. B. View from cross-section. C. Surface view of neurula with beginning neural folds. D. Cross-section through neurula as in C. From Spemann, 1936, after drawings by V. Hamburger and B. Mayer.

At the end of gastrulation, three layers can be distinguished, which are designated germ layers. The outer lining is the ectoderm; the middle layer, the archenteron roof, is the mesoderm; and the primitive intestine is the endoderm. The significance of the germ layers is that each gives rise to a particular set of structures and organs; the germ layer derivatives are characteristic of all vertebrates.

Gastrulation is followed by neurulation, the formation of the nervous system (Fig. 2–9). A pear-shaped area of the dorsal ectoderm thickens, forming the neural plate. Neural folds arise along the circumference of the neural plate, grow upward, and bend toward the midline, where they eventually fuse. Thus the neural tube is formed. The wider anterior part represents the anlage of the brain; the posterior part will form the spinal cord. The brain vesicle is divided first into forebrain, midbrain, and hindbrain. The forebrain is subdivided further into the telencephalon and diencephalon, the hindbrain into cerebellum and rhombencephalon or medulla oblongata. The eye vesicles are protrusions of the forebrain; the ear vesicles are derived from local thickenings of the ectoderm lateral to the hindbrain, the otic placodes.

The gastrula and the neurula are spherical. After the completion of neurulation, the embryo elongates and a tail bud is formed (Fig. 2–10). I shall describe briefly the mesodermal structures which are differentiated in the tail bud stage; they will be referred to frequently in the following discussion. A median dorsal strip of mesoderm underneath the neural tube becomes segregated; it is a transient skeletal rod called the notochord. The mesoderm on both sides of the notochord

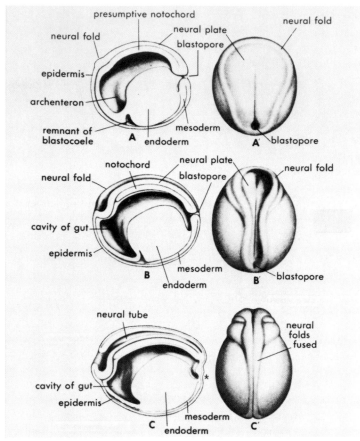

Fig. 2–9. Schema of neurulation in frog embryo. A–C. Views from median plane. A′–C′. Surface views. From Balinsky, 1978.

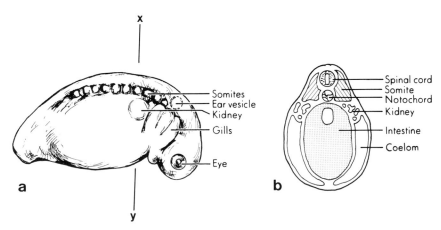

Fig. 2–10. Tail bud stage of salamander. a. Side view. b. Cross-section in X—Y plane, showing the axial organs. Original drawing.

becomes segregated into blocks of tissue, the somites. They differentiate in ante-rior-posterior sequence; in the early tail bud stage only eight to ten pairs of somites are formed. The somites give rise to the trunk musculature, the vertebrae, and the deep layer of the skin, the dermis. The lateral and ventral part of the mesoderm mantle, the so-called lateral plate, splits into two layers. They become the inner and outer wall of the coelom, or body cavity, which expands between them. A strip of mesoderm between the somites and the lateral place differentiates to a set of convoluted kidney tubules (pronephros). They open into a duct which grows backward toward the anus. (The formation of the tail will be discussed in the last chapter).

The constriction experiments showed that the capacity for the formation of duplications, that is, regulation, is lost at the end of gastrulation. Hence, the deci-sive determinative events for axial differentiation occur during the gastrulation process. This crucial observation was the incentive for a set of experiments which signified a breakthrough in experimental embryology. The new tools for micro-surgery made it possible to extend the longstanding practice of transplantation to very early stages of development.

Spemann's plan was to make a comprehensive test of the state of determination of the different regions of the early gastrula. He proceeded by transplanting small parts of the gastrula to different regions of an embryo of the same stage. Would the transplant follow its own intrinsic instruction, disregarding its new environ-ment? Or, on the contrary, would it conform to its new environment, thus dem-onstrating its uncommitted state? The design of the experiment was ingenious in that transplants were exchanged in such a way that the transplant taken from one gastrula was fitted in the hole from which the other transplant had been taken. In this way, each transplant would not only answer this question, but at the same time indicate the site from which the other had originated. In the absence of pre-cise maps of the surface of the gastrula—the vital-staining maps of Walther Vogt were not yet available—he had to make informed guesses about the location of different structures. Fortunately, the blastopore provides some points of refer-ence. As was mentioned, it demarcates the posterior end of the embryo. The region above it is dorsal, and the plane that bisects it is the median plane of the embryo.

Two further technical prerequisites had to be met. It was desirable that the transplants be of equal size and shape. This was accomplished by using the micro-pipette for the excision of the transplants. The selected surface area was sucked in and cut off at the open end of the pipette with a fine lancet. The transplant was then transferred to the other gastrula, on which the same operation had been per-formed previously. The transplant was fitted in the proper hole and held in place with a glass bridge until it was healed in; this took no longer than an hour. It was also desirable to keep track of the transplants as long as possible. This was achieved by taking advantage of the differences in pigmentation of the eggs of the common salamander, *Triturus taeniatus,* which ranged from light to dark brown. The transplants on the surface remain recognizable through the neural plate stage and even through the early tail bud stage; that is, their position in the neural plate and epidermis, respectively, can be identified. However, they are not discernible in microscopic sections; hence, the fate of invaginated material was difficult to determine. This was a serious handicap which delayed further analysis until Spemann overcame it by adopting Harrison's method of transplantation between different species (heteroplastic transplantation). I have described the experimental procedure in detail, because it was also used later in the organizer experiment.

The analysis began with the simplest test: the exchange between prospective neural plate and prospective epidermis in the early gastrula. The results were clear-cut. All transplants developed in accordance with their new environment: prospective anterior or middle neural plate material became typical epidermis, and prospective epidermis formed typical neural structures, such as specific brain parts and eyes. In this process, the originally circular transplants changed shape: those in the epidermis enlarged and formed patches of different contours; those in the anterior neural plate remained roundish; but those in the middle and posterior neural plate were transformed into narrow longitudinal strips, indicating a conspicuous stretching along the main axis during neurulation. In one case (Fig. 2–11), a transplant inserted halfway between the blastopore and the animal pole formed a narrow lightly pigmented strip in the neural plate which extended to the posterior end. Spemann concluded, "Hence two regions can be distinguished in the [prospective] neural plate: an anterior region which does not undergo essential movements from the beginning to the end of gastrulation, and a posterior region which narrows and, at the same time, stretches enormously" (1918, p. 517). An exchange between the same regions at a stage when the neural plate just becomes visible showed that the different structures had by then become irreversibly committed; prospective neural tissue differentiated into brain parts; the prospective eye region of the anterior neural plate differentiated into an optic cup (Fig. 2–12); and prospective epidermis formed a patch of epidermis in the nervous system. These experiments were mentioned only briefly in the publication of 1918 and described in detail later (1919).

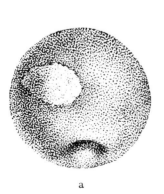
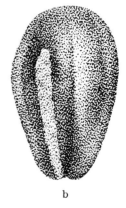

a   b

Fig. 2–11. Homeoplastic transplantation in *Triturus taeniatus.* a. Lightly pigmented implant derived from prospective flank epidermis of the donor embryo. b. Same embryo in early neurula stage. The transplant forms a narrow strip in the neural plate. From Spemann, 1918.

From these data the general conclusion was drawn that the determination of the neural plate occurs during gastrulation and becomes irreversible when the neural plate makes its first appearance. This result came as no great surprise, for it had been implicit in the constriction experiments. However, the great superiority of the transplantation method over constriction revealed itself in the next experiment that Spemann undertook: the transplantation of the upper blastoporal lip of the early gastrula to the flank of another embryo of the same stage. This was a momentous step. The experiment was actually an early version of the organizer experiment. It differed from it in that it was done between embryos of the same species and thus lacked a permanent marker. This shortcoming made the interpretation of the results equivocal and concealed their true significance.

The transplant, representing the invaginating zone, consisted of an outer and an inner layer; they were continuous around the blastoporal lip. The experiment was apparently not done on a large scale. Only three cases were described and they had been sacrificed in early tail bud stages. In two of them the embryos showed a long narrow ridge on their flanks which was identified in sections as a second neural tube underlain by a notochord flanked by two rows of somites; obviously, an incomplete axial system of the trunk region had differentiated. In the third case, the transplant had been placed close to the main axis of the host and had differentiated as a short side branch of the trunk axial system of the host.

How did Spemann interpret the experimental data? He recognized immediately the most significant point: the exchange experiments on the *early* gastrula already described had shown that those regions between the blastopore and the animal pole that were destined to become anterior and middle neural plate were still

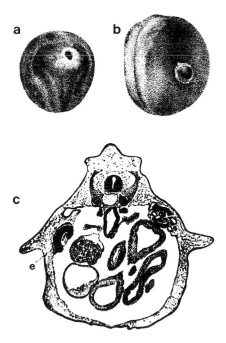

Fig. 2-12. a,b. Transplantation of the eye-forming region in the anterior neural plate of a frog embryo to the flank of another embryo. c. Cross-section of the larval stage at the level of the transplant. e. Eye, differentiated from the transplant. From Spemann, 1919.

uncommitted, since they had formed epidermis in the epidermal environment. In contrast, the region adjacent to the blastopore when transplanted to a similar position had given rise to a neural tube, among other structures. Hence, the state of commitment of the blastoporal region was more advanced than that of more anterior regions. Spemann concluded that the determination of the neural plate begins at the blastopore and proceeds from there in the anterior direction. This inference led to the designation of the blastoporal lip region as the *differentiation center* (1918, p. 530). A year later, the term *organization center* was coined and used synonymously with differentiation center (1919, p. 584).

The concept of the differentiation center followed logically from the transplantation experiments. Yet for Spemann, the discovery was not an unforeseen revelation. It has been said that it takes a prepared mind to make an extraordinary discovery. Spemann's mind had been prepared for two decades; the preconception of the differentiation center dates back to one of his earliest investigations (1898). He refers to it in his book:

> The notion of determination progressing from cell to cell, which led to my conception of the "differentiation center," had been suggested by facts encountered at the very beginning of my research. I had studied the development of the middle ear in anuran amphibians [frogs and toads] in connection with the transformation of the chondrocranium [cartilaginous head skeleton] during metamorphosis. As is known, a considerable part of the chondrocranium is dissolved and new cartilage is formed in connection with the residual cartilage. Then I conceived the concepts of "expansive" and "appositional" growth. (1938, p. 142)

A few years after the original observation, he gave the following definitions: "The growth of an anlage can occur either by growth and division of cells, that is, expansive growth, or by a process whereby already existing but indifferent cells are incorporated in the anlage; this is growth by addition [Andifferenzierung]" (1903a, p. 606). In the discussion in the publication of 1918 the term "appositional growth" is substituted for the term "Andifferenzierung" (1918, p. 541), and much later, in his book Spemann used the term "assimilative induction" (1938, p. 163). He acknowledged his longstanding familiarity with assimilative growth in the following statement: "This notion has always remained in the background of my consciousness, so that I interpreted the first case of assimilative growth which I encountered, in this sense" (1931a, p. 511). What had been a hypothetical construct for so many years now became concrete reality in the determination of axial mesoderm and neural plate. Spemann could not have wished for a more propitious affirmation of his early intuition; it was tied to another favorite theme of his, the determination of the axial organs.

The designation of the blastoporal lip as the differentiation center for the axial organs is valid only in a very general way; it becomes ambiguous and, in fact, it became the seed of a profound misunderstanding when applied specifically to the neural plate. The problem of neural plate formation had intrigued Spemann from early on. When engaged in the constriction experiments, he had been impressed by the consistency with which in all anterior duplications the extent of the duplicated brain had conformed precisely with the extent of the subjacent mesoderm, the so-called archenteron roof. This and similar observations had suggested to him that a causal relation might exist between the mesoderm and neural plate formation. "It is conceivable that the neural plate is induced by the archenteron [mesoderm]" (1903a, p. 616). This is the first inkling of what became an important deduction from the organizer experiment. It should be kept in mind that the idea of neural plate induction by the subjacent mesoderm dates back to 1903! But now an alternative interpretation of neural formation suggested itself: beginning at a neural differentiation center in the upper layer of the blastoporal lip, neural differentiation might spread forward in the ectoderm by assimilative induction. The two alternatives are stated succinctly in the lecture of 1919:

> It appears that a piece near the upper blastopore has progressed further in its development and is more firmly determined than a piece at a greater distance from the blastopore. In other words, the determination seems to progress from the blastopore in the anterior direction. How this occurs cannot be stated with certainty. The two experiments—transplantation of a piece of ectoderm directly above the blastopore and a piece [of ectoderm] located more anteriorly—differ from each other not only in their [topographic] position but also in that the anterior piece consists of pure ectoderm, whereas the one near the blastopore is so closely attached to the underlying cell layers of the meso- and endodermal germ layer that it has not been possible so far to separate them with certainty and to transplant pure ectoderm. It could be that the deeper cell layers cause the ectoderm to pursue its normal differentiation [to neural plate]. Therefore, the progression of determination could proceed entirely in the ectoderm, whereby the cells which are already determined initiate the same differentiation in the cells located in front of them. Or the progressing determination [of the neural plate] could be mediated by the underlying mesodermal and endodermal cell layers, which, during gastrulation, advance forward beneath the ectoderm. In both instances, we are dealing with the growth of an anlage which occurs not by multiplication and growth of its cellular elements [expansive growth] but by incorporation of new, previously indifferent elements [appositional growth]. (1919, pp. 584–585)

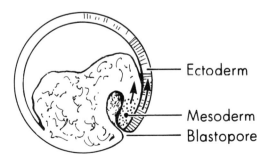

Ectoderm

Mesoderm

Blastopore

Fig. 2–13. Spemann's interpretation of the fate of the two layers of the upper blastoporal lip. The outer layer is mistakenly identified as prospective neural plate (ectoderm) instead of mesoderm. *Arrows* indicate his notion of the progression of determination from the organization center at the upper blastoporal lip. Original drawing.

Spemann recognized that the available data did not permit a decision between the two alternatives, that is, determination of the neural plate by the underlying mesoderm or by a neural organization center located in the upper layer of the blastoporal lip. What he did not realize was that his position was seriously flawed by his erroneous interpretation of the nature of the blastoporal lip. He mistook its upper layer for neural ectoderm, whereas in reality it is mesoderm destined to invaginate in the course of gastrulation (Fig. 2–13). In the quotation above he refers to "a piece of ectoderm directly above the blastopore." This is not an isolated oversight. The section dealing with the transplantation of the upper blastoporal lip in the publication of 1918 begins with the following sentence: "In several cases, a piece very close to the upper blastoporal lip of an early gastrula was transplanted to the prospective epidermis of another embryo. In these cases the *ectoderm of the piece* developed to neural tube; hence it behaved differently from the more anterior ectoderm of the same stage" (1918, p. 477; italics mine). I found numerous additional references to the "ectoderm near the blastopore." At one point Spemann makes the unequivocal statement, "It is possible that the center for the differentiation of the neural plate, at the beginning of gastrulation, is contained in the ectodermal part of the blastoporal lip and that differentiation spreads forward purely in ectoderm" (1918, p. 532). In other words, the idea that the upper layer of the blastoporal lip is ectodermal and that the neural plate of the transplant would develop by self-differentiation of its upper layer had firmly taken hold of his mind.

A simple reflection on the mechanism of gastrulation, however, leaves no doubt that this assumption cannot be correct. Even granted that gastrulation was poorly understood at the time, and that hindsight, benefitting from Vogt's beautiful analysis of gastrulation movements by vital staining, gives us an undue advantage, the mesodermal nature of the upper layer of the blastoporal lip should have been obvious to Spemann. It is not difficult to envisage that cells located above the blastopore at the early phase of gastrulation will invaginate in the subsequent phase and disappear inside to form part of the mesoderm mantle. One cannot even plead that, for once, Spemann's otherwise unerring visual imagination failed him, and that this was an understandable oversight on his part. To my surprise, I found in another part of the publication of 1918 the description of an experiment that had given unmistakable evidence that the upper layer of the blastopore is, indeed, mesodermal, and at that point he had accepted this proof without reservation. I refer to a case in the series of exchange transplantations between prospective neural plate and prospective epidermis in the early gastrula which has been mentioned above. A circular piece of prospective epidermis from a lightly pigmented embryo had been inserted at some distance above the blastopore of a darker early gastrula. A day later, the whitish transplant was visible as a long narrow strip on the left side of the neural plate of the dark embryo (Fig. 2–11). The important point is that it extended all the way to the posterior end of the neural plate. This implies that the region between the posterior end of the transplant (in its original position) and the blastopore must have invaginated, or, in other words, that both layers of Spemann's transplants of the upper lip to the flank were mesodermal in nature. Spemann acknowledges this in his comment on this case:

"The round white transplant must have been transformed into a long narrow strip under the influence of its new environment. It follows that the material extending from its posterior edge to the blastopore *has been invaginated during gastrulation*" (1918, pp. 516–517, italics mine). This flash of insight, however, was ephemeral. Far from revising his erroneous notion of the ectodermal nature of the upper layer of the blastoporal lip, he continued to insist on its mistaken identity. In the lengthy discussion in the publication of 1918, the single acknowledgment of the true nature of the blastoporal lip was outweighed by the numerous statements to the contrary that I have mentioned. One is puzzled not just by the inconsistency but by the undisguised preference for the incorrect interpretation.

How can one account for Spemann's tenacious insistence on the erroneous interpretation of neural plate determination? Of course, it is impossible to divine the true motives for his preference, but I shall venture a guess. The notion of neural plate determination within a single layer, beginning at the differentiation center, seemed to give reality to the concept of "appositional growth" which was envisioned in 1898. It was probably his first truly original idea and, as such, cherished and reinforced as a measure of self-affirmation. It remained a dominant theme of his thinking for decades. This is borne out by the recurrence of references to "appositional growth" in many of his publications (1903a, p. 606; 1919, p. 585; 1938, p. 143, Spemann and H. Mangold, 1924, pp. 628, 633, in addition to those in the publication of 1918). At any rate, this model so dominated the notion of neural determination by the underlying mesoderm that the latter was repressed. It was even underplayed in the discussion of the organizer experiment, although, ironically, neural induction by mesoderm later became the hallmark of organizer activity.

In dwelling at some length on Spemann's misconception, I certainly do not intend to show once more that great minds can err. Rather I consider his curious prejudice as another interesting illustration of the old truth that the exploring scientific mind, when arriving at a crossroad, makes decisions on the basis of both rational and not entirely rational considerations. In research, one is guided by some frame of reference that may be called a working hypothesis, but also by hunches, preconceptions, and even strong personal preferences that may be rooted in very deep strata of the personality. Science is not all that objective; this has been pointed out many times. And the steps from intuition to preconception, and from there to inflexible tenets, are sometimes short.

Spemann's error had some serious consequences. In all probability, it accounted for his failure to undertake the blastoporal lip transplantations on a larger scale in 1916. If, as he believed, the two layers of the transplant simply proceeded by self-differentiation, independently of each other, then the outcome was not very exciting, because self-differentiation was a familiar phenomenon. Probably for the same reason, the experiment was not repeated in the following year, when he had improved the technique by transplanting from unpigmented to pigmented embryos. As a result, the discovery of the organizer was delayed by four years!

I have dwelled at length on an experiment which involved only three cases and comprised only a small part of the lengthy publication of 1918. The reason is obvious: the discovery of the differentiation center, though its significance was poorly understood and though it was misinterpreted in part, was a crucial step on the way to the discovery of the organizer and one of Spemann's major achievements. But I admit that I was also intrigued by the denouement of the strange case of mistaken identity which had escaped me, although I had studied this paper many times in the past.

The other two experiments reported in 1918 continued the analysis of the state of determination of the early gastrula by manipulation of gastrula halves rather than by transplantation of small pieces. In the first, the gastrula was transected by a horizontal cut above the blastopore, using glass needles, and the two halves were fused again, after rotation of the upper half by 90° or 180°. The operation is more difficult than transplantation of small pieces. The cut surfaces have to fit tightly and the two halves have to be held together by glass bridges for about an hour. The remarkable healing power of these early embryos has to combine with the skill of the experimenter to assure success. Again embryos with considerable differences in pigmentation were chosen in order to be able to trace the contributions of each part at least to the neurula stage. The embryos were completely normal; two of them were raised to advanced larval stages.

The rotation experiment supplements and extends the transplantation of small pieces. It gives impressive evidence for the versatility of the ectoderm of the early gastrula. The exchange of small pieces between prospective ectodermal and neural areas had shown already that their fate could be influenced by their new environment. In the rotation experiments, the differentiation of large parts of the ectoderm was directed along pathways that were different from their normal course. In the 180° rotations, the part that would normally have formed belly ectoderm now formed neural plate, and in the 90° rotations, the disposition of materials was also different from their normal fate. One gets the impression that the ectoderm at this stage is entirely uncommitted or indifferent. Spemann, however, did consider the possibility that prospective neural and epidermal ectoderm were already predisposed toward their respective fates and that these tendencies were overridden. While this seems plausible in the case of small transplants which were completely surrounded and, so to speak, overwhelmed by their new environment, the change of fate of the entire upper hemisphere speaks more in favor of a lack of commitment. Since the extraordinary plasticity of the gastrula ectoderm plays a key role in the experiments that followed the organizer experiment, it is of historical interest that the rotation experiment was the first in which this point became manifest.

This experiment also reinforced the notion of a differentiation center in the region of the blastopore, since it is obvious that the blastopore in the lower hemisphere had imposed its axial orientation and plane of symmetry on the neural plate in the upper rotated hemisphere. As far as the mechanism of neural plate determination is concerned, all that can be said is that if Spemann had fully realized and accepted the notion of the mesodermal nature of the blastopore, the outcome of the rotation experiment would have convinced him then and there of the reality of neural plate induction by the underlying mesoderm. As it was, the question remained unresolved in his mind.

The third experiment reported in 1918 is important because it prepared the way for a comprehensive theory of the origin of double monsters in animals and man, a theme which, we remember, was the starting point of Spemann's analysis of the determination of axial organs. This time, two gastrulae were split in half in the median plane of the blastopore; then, the two right halves and the two left halves were fused together. By rotation, the two half-blastopores were placed at some distance from each other. Each half-blastopore then instigated the adjacent ectoderm to participate in the invagination, and the end result was the formation of two complete embryos at opposite sides. This was a new way of producing duplications. Spemann then modified the procedure and combined two gastrula halves, each containing a whole blastopore. This experimental design had a greater potential for further analysis than the previous one. I shall deal with these experiments presently.

Teratologists interested in the origin of naturally occurring ("spontaneous") duplications had wondered whether they are caused by the splitting of a single primordium or the fusion of two primordia that had formed in a single egg. Spemann's constriction experiments had shown that in amphibians duplications can be obtained by fission. But he had found that they were invariably anterior duplications, and he had given a satisfactory explanation of this fact in terms of the anterior direction of invagination. Posterior duplications, however, cannot be explained in this way, and this holds true also for a third major type, the *duplicitas cruciata*. This term was introduced by Spemann as a simplification of what is known in the cumbersome terminology of teratology as *cephalothoracopagus*. It is a most bizarre and rare monster in which two heads facing in opposite directions continue in two separate trunks and pairs of arms and legs whose plane of symmetry is perpendicular to that of the heads. Occasional asymmetries in the faces suggest that the latter are actually composites, and that the left half of each face is derived from one primary unit and the right half from the other.

Spemann made a major contribution to the theory of duplications by basing his explanation of their origin on the notion of the newly discovered differentiation center. He considered all three types of duplications as variations of one theme: he assumed that, atypically, some eggs form two differentiation centers instead of one and that the different types of duplications can be explained in terms of different positions of the two centers with respect to each other:

> If it were possible to produce embryos with two differentiation centers, one would expect that double monsters of different types would result. The types would depend on the position which the centers occupied with respect to each other. If their median planes diverged in the anterior direction, then animals with anterior duplications would originate. If they converged, animals with a duplication of the posterior ends would result. If the streams of differentiation collided head-on, the fission and fusion of the anterior ends would form a cross, and those peculiar monsters that could be called 'duplicitas cruciata' would be formed. (1919, p. 585)

Incidentally, one notices that in *duplicitas cruciata* fission and fusion are no longer considered as alternatives by Spemann but as processes occurring side by side.

In his experiments, Spemann followed precisely the plan of combining two half-gastrulae, each containing a whole differentiation center, and the outcome confirmed the predictions, and thus the theory, in every detail (Fig. 2–14). He managed to create *duplicitas cruciata* in salamanders in the following way: The upper hemispheres of two gastrulae were cut off at some distance from the blastopore, as in the rotation experiment. The two lower parts were then fused, the upper blastoporal lips facing each other. Since the distance between the two blastopores was shorter than the length of the future embryos, the invaginated mesodermal sheets met in head-on collision along the line of the suture; they deflected each other and at the same time fused together along a plane perpendicular to that of their original plane of invagination. In this way, heads and trunks formed a cross; each trunk and its appendages were derived *in toto* from one of the original half-gastrulae; but the heads were composites: one half of each head was derived from one of the original components and the other half from the other.

Anterior duplications were produced by fusing two half-gastrulae in such a way that the blastopores were adjacent to each other, but their median planes *diverged* at an acute angle. Thus the two invaginating mesoderm layers remained separate and induced two neural plates, whereas the posterior regions remained uniform.

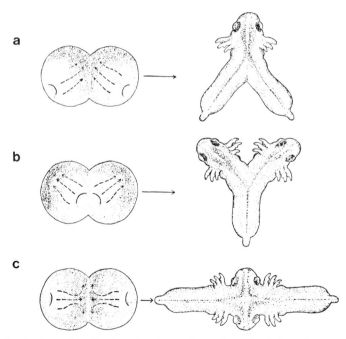

Fig. 2–14. Production of duplications in salamander embryos by fusion of two gastrulae whose upper or lower caps had been removed. a. Posterior duplication produced by fusion of gastrulae in which the direction of invagination converges *(broken arrows)*. b. Anterior duplications. The direction of invagination diverges *(broken arrows)*. c. *Duplicitas cruciata*. Invagination occurs in opposite directions, leading to the diversion of invagination in a plane perpendicular to the original direction of invagination *(broken arrows)*. From J. Holtfreter, 1951.

Posterior duplications were obtained by a modification of the *duplicitas cruciata* experiment. The upper hemispheres of two gastrulae were cut off obliquely, and the cut surfaces of the lower parts were fused together in such a way that the two blastopores were again located at opposite sides, but instead of facing each other, their median planes *converged* at an acute angle. Thus, the invaginating mesoderm layers merged and induced jointly a single anterior neural plate, whereas the posterior ends remained separate and, in fact, diverged. Apparently, these experiments had not been completed when the manuscript for the publication of 1918 was prepared. Descriptions of the duplications were presented with many illustrations in the Inaugural Lecture of 1919.

Despite its shortcomings, Spemann's publication of 1918 can be acclaimed as a major breakthrough in the attainment of his primary objective: to understand the mode of determination of the axial organs. As we have seen, the constriction experiments had identified the gastrula as the critical stage at which determination occurred. But the severe limitations of the constriction method had been a block to further progress. This obstacle had been overcome by the ingenious application of the glass needle and micropipette technique to early gastrula stages. The transplantation, rotation, and fusion experiments had led to the discovery of the differentiation or organization center and given it a solid foundation by the multifaceted approach. It is of historical interest that the first tangible benefit from this discovery was a unified theory of the origin of different types of duplications and its experimental validation. The next logical step in the analysis was to remove the second major stumbling block, the lack of a permanent marker. This was achieved by the application of the method of heteroplastic transplantation to early gastrulae, carried out in 1917 and published in 1921.

If one remembers that the years 1918 and 1921 also saw the publication of two classical experiments of Harrison, and that Spemann's paper of 1921 reported in a postscript the discovery of the organizer, then one can truly say that these years marked the pinnacle of experimental embryology. Harrison's "Experiments on the development of the forelimb of *Amblystoma,* a self-differentiating, equipotential system" (1918) were a fundamental advance in the analysis of the regulative capacities of developmental systems. His experiments "On relations of symmetry in transplanted limbs" (1921) provided an ingenious solution of the problem of determination of bilateral symmetry in vertebrates. As such, they shed light on a major issue in the determination of vertebrate organization and thus converged on Spemann's focus of interest. Spemann had paid only passing attention to this facet, in a brief communication on determination of symmetry in otocysts (1910). Thus the achievements of Harrison and Spemann complemented each other in providing major advances in our understanding of fundamental aspects of vertebrate development.

Heteroplastic Transplantation (1917–1921)

Spemann's experiments on heteroplastic (interspecific) transplantations followed those on homeoplastic transplantation within a year; the latter were done in 1916 and the former in 1917. The first results of the heteroplastic experiments were mentioned and illustrated in the Inaugural Lecture of 1919 and fully published in 1921 under the title "The production of animal chimaeras by heteroplastic embryonic transplantation between *Triton cristatus* and *taeniatus*." Chimaera is the name of a monster in Greek mythology that combines a lion's head and a goat's body with a serpent's tail. The term found its way into biology when botanists adopted it early in this century for experiments in which shoots of one species were grafted onto a plant from another species. New growth was then elicited at the border of the two components. The ensuing outgrowth was composed of layers derived from both components and called a chimaera (Winkler, 1907). A few years earlier, in 1903, Harrison had produced animal chimaeras, without using the term. In a classical experiment, he had combined the head of a tail bud stage of a darkly pigmented frog *(Rana palustris)* with the trunk and tail bud of a lightly pigmented frog *(Rana sylvatica)*. He observed the deposition of several rows of dark lateral line sense organs on the light trunk and tail of the *sylvatica* tadpole and was able to trace their origin to an epidermal thickening in the hindbrain region, the postotic placode (Harrison, 1903). In his book, Spemann remarks, "[from this experiment] I have learned more than from almost any other investigation, not only for technique, but also for advancing analysis" (1938, p. 131). The innovation of Spemann's experiment was the extension of the method of heteroplastic transplantation to early gastrula stages. Its immediate purpose was to provide a permanent marker. The German fauna cooperated propitiously with this enterprise by providing him with the unpigmented white eggs of *Triturus cristatus* and the brownish eggs of *Triturus taeniatus*. In the past he had used only the latter, because they are more hardy than those of *cristatus,* when all membranes are removed. But the potential of the method of heteroplastic transplantation goes far beyond its usefulness for providing markers. It is a powerful tool for the analysis of tissue interactions in development, as will be discussed below.

The design of the experiment was not novel: it was a repetition of the experiment of 1918 in which small transplants were exchanged between prospective neural plate and prospective epidermis of the early gastrula. The experiments were done again with the micropipette, which permitted a neat exchange of pieces of identical size and shape; again, each transplant revealed the site from which its partner had been taken. The novel aspect was that transplant and host tissue could be identified on the microscopic level.

It is characteristic of Spemann's style that in his publications he describes and illustrates only a few representative cases from which he then draws far-reaching conclusions. This format is based on the accepted view that in experimental embryology a single unequivocally positive case in a well-designed experiment can give a conclusive result. Of course, he and everybody else repeated the experiments, until he had convinced himself of the consistency of the results: but it became customary to select the best cases and to describe them in great detail. Since some authors indulged in presenting many cases, the publications of this period often became very lengthy and tedious (those of the present author not excluded). As far as I can see, Spemann never tabulated his results. But there were also practical limitations to the collection of large numbers of experimental cases.

Spemann states apologetically, "The considerable technical difficulty of the operation and the very great mortality of the larvae may explain the fragmentary character of my communication" (1921, p. 565). I myself have never worked on very early stages of amphibian embryos, but I remember vividly from my Freiburg days the frustrations of those around me who did.

At any rate, in the paper under discussion, altogether nine individual cases were mentioned, and three of them were used to exemplify and illustrate specific results. The impressive pictures illustrating the first case have been reproduced in most books and publications dealing wtih experimental embryology: they present the large *cristatus* gastrula with a round dark *taeniatus* implant, side by side with the small *taeniatus* gastrula with its white transplant, and the neural plate stage of the *taeniatus* partner with a somewhat elongated white area in the left anterior plate (Fig. 2–15) and a cross-section of the head of the same embryo a few days later, showing a very distinct unpigmented sector of the forebrain and an unpigmented eye on one side (Fig. 2–16a). The borders between donor and host tissue can be traced down to individual cells. The smooth incorporation of tissue from the foreign species is very impressive indeed. The tissue that was originally destined to become epidermis had undergone all morphogenetic processes in conformity with its new environment. Yet the species characteristics of *cristatus* are expressed in the greater thickness of the brain wall, due to the larger size of individual cells. The *cristatus* partner of this embryo has not enjoyed the same publicity, although it deserved it. The transplant, which was destined to become forebrain and eye, now forms a dark patch of flank epidermis (Fig. 2–16b). The embryo was raised to an advanced stage characterized by the appearance of gills.

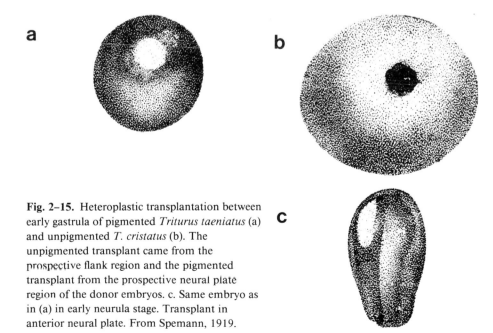

Fig. 2–15. Heteroplastic transplantation between early gastrula of pigmented *Triturus taeniatus* (a) and unpigmented *T. cristatus* (b). The unpigmented transplant came from the prospective flank region and the pigmented transplant from the prospective neural plate region of the donor embryos. c. Same embryo as in (a) in early neurula stage. Transplant in anterior neural plate. From Spemann, 1919.

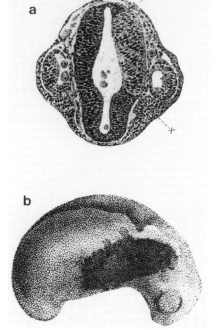

Fig. 2-16. a. Embryo shown in Fig. 2-15c in tail bud stage. Cross-section through brain. Note unpigmented eye and part of brain wall on right side (sector X—X) formed by transplant. b. Embryo shown in Figure 2-15b in tail bud stage. Note pigmented transplant on right flank covering gill region. From Spemann, 1919.

The transplant covers the gill region on one side. The single most interesting feature of all these experiments is the presence of three distinct buds, the incipient gills, on the transplant side; in contrast, these buds are not yet discernible on the other side. Here the gill region is less advanced: it is demarcated by a smooth swelling. The conclusion was drawn that the epidermal covering plays a significant role in regulating the speed of development of the underlying mesodermal and endodermal components of the gills. This was corroborated by the presence of an aortic blood vessel in the chimaeric gill: its counterpart was not yet formed on the other side.

In the second case, a *cristatus* transplant has formed an inner ear vesicle (otocyst) in a *taeniatus* host embryo. The sections show not only the absence of pigment but a distinct difference in the shape of the vesicle; this is again the manifestation of a species-specific characteristic.

A critical assessment of the actual results of the experiment leads to the conclusion that, apart from the introduction of a permanent marker, they were very modest; indeed little advance was made in the analysis of early development. But the publication of 1921 must be evaluated in terms of the long-range potential of the method; from this vantage point, it ranks as one of Spemann's major contributions. First and foremost, it was the indispensable stepping-stone to the organizer experiment. Second, it opened a new approach to the analysis of tissue interactions in development which is a dominant theme in experimental embryology. Most organs are composites of tissues that originate in different regions or belong to different germ layers and are brought together by morphogenetic movements.

One example, the gills, in which all three germ layers are involved, was mentioned. The observation that in this instance the ectodermal covering takes the lead in instigating the time of outgrowth of the gills illustrates the kind of questions that can be answered. By exploiting morphological species differences, one can gain deeper insights in tissue interactions. Spemann gives an example of another experiment that would be feasible. The legs, and particularly the toes, of the two *Triturus* species show distinct differences in their development: those of *T. taeniatus* remain rather short and stubby, while the toes of *T. cristatus* become very long and very slender. One can build chimaeric limb buds with the mesodermal core of one species and the epidermal covering of the other, and one can then "examine whether the legs which are covered by epidermis of a different species are determined in their shape by the formative tendencies of the epidermis or of the inner structures or both" (1921, p. 562). This experiment was done later by Spemann's student Eckhard Rotmann (1931, 1933). He found that the mesodermal component expresses its species-specific characteristics and that the epidermis is capable of adapting itself to the developmental design of its foreign partner. At the same time, Harrison and his students took advantage of the considerable size differences between the American salamanders, *Ambystoma tigrinum* and *A. punctatum,* and their embryos, for an analysis of growth regulation. Different organs of the two species, such as eyes and limbs, have different growth rates and sizes. Take, for instance, the case of lens induction. How would the lens epithelium of the small species, *A. punctatum,* interact with the optic vesicle of the large species, *A. tigrinum?* Harrison showed by meticulous measurements a mutual adjustment of the two components. On the other hand, when entire limb primordia are exchanged between the two species, they retain their intrinsic growth rates; the same holds for reciprocal transplants of whole eyes. A succinct discussion of the application of heteroplastic transplantations to quantitative problems of growth may be found in Harrison's Harvey Lecture (1935). I mention his work in this context to illustrate the versatility of the method. Actually, as was pointed out earlier, he had used it long before Spemann.

Spemann was never particularly interested in quantitative aspects of development. His major concern was tissue interactions, and more specifically, the mechanism of embryonic induction. What he considered as one of the fundamental problems is best stated in his own words:

> Is the determinative effect of the environment on the relatively indifferent embryonic material a directly determinative [instructive] one or only a releasing one; [auslösend]? The overwhelming probability speaks in favor of the latter. The ectoderm, which has the choice between neural plate and epidermis, is directed to one or the other only in a general way, and it pursues its further differentiation according to its inherent disposition. The key word given is, so to speak, quite general: 'neural plate' or 'epidermis.' The execution in detail occurs according to the formula that is contained in the genetic constitution of the species. (1921, p. 567)

Continuing this line of thought, he had one of his boldest and most ingenious inspirations:

If it should be possible to exchange transplants between a urodele and an anuran, then one could replace the ectoderm of the mouth region of the former with that of the latter. Would the larva develop teeth [characteristic of the salamander] or horny jaws [characteristic of the frog tadpole but nonexistent in the salamander larva]? If the latter is the case, then the key word would be very general 'mouth armament' and the ectoderm would furnish the kind of armament appropriate to it. (1921, p. 567)

At that time, Spemann saw little hope that this "thought experiment," as he called it, would be feasible. But the dream became reality a decade later. Oscar Schotté succeeded in what had seemed to be an unrealistic project. He transplanted ventral ectoderm of the frog gastrula to the mouth region of a salamander gastrula and *vice versa,* and he managed to raise some of the experimental embryos to larval stages. He obtained salamander larvae with frog-type horny jaws and suckers (Fig. 2–17) and frog tadpoles with balancers (a pair of balancing rods on the head of salamanders, but not of tadpoles) and dentine teeth. (Spemann and Schotté, 1932; Spemann, 1938, p. 358ff.) The results and conclusions drawn from this experiment were exactly as predicted by Spemann. This was undoubtedly one of the most spectacular experiments in experimental embryology. Considering that the two orders of amphibians, anurans and urodeles, have split off from a common ancestor in the Paleozoic, 350 million years ago, it is remarkable that their embryos have retained the inductive capacity to elicit typical mouth structures in each other's ectoderm, including structures which are not in the genetic repertory of the host species. The inductors were identified later as head mesectoderm and endoderm.

Oscar Schotté had joined Spemann's laboratory in 1928 as a research fellow. He came from Geneva, where he had received his Ph.D. in the department of E. Guy-

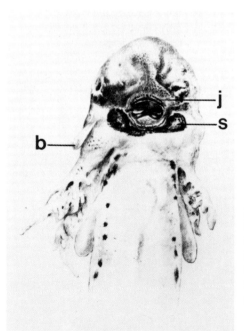

Fig. 2–17. Xenoplastic transplantation. Salamander larva with mouth implements of frog larva: horny jaw (j) and suckers (s). The balancer (b) is typical of salamander larvae. In the gastrula stage, belly ectoderm of a frog embryo had been transplanted to the head region of a salamander embryo. From Spemann, 1936.

énot (a recognized authority on amphibian limb regeneration) and advanced to an assistant professorship. He had already published extensively on regeneration, and this topic remained his lifelong primary interest. He was a welcome addition to the Freiburg laboratory; he brought to the rather serious-minded German group a touch of French vivaciousness (he was half French). He had grown up in Poland and Russia, and through many travels had acquired a cosmopolitan outlook. I remember his animated disputes, in Russian, with another visitor in the laboratory, Dr. George Schmidt, from Moscow—which had decidedly political overtones. (One day, Schmidt suggested to Spemann that the organizer might be compared to a labor organizer, since both influence relatively indifferent masses. Spemann, predictably, rejected this as an affront to the embryo.) Schotté was also a great storyteller, and he convinced me that one can be debonair and a very good scientist. His entrepreneurial spirit made him ideally suited for the undertaking of the daring experiment which Spemann himself had never tried. I remember vividly our excitement over his first results, in the spring of 1931, reminiscent of our thrill a decade earlier, when Hilde Proescholdt had shown us the first organizer-induced embryo. But the scenario was quite different. Schotté had pledged that he would not shave during his exertions; his beard grew, until one morning he appeared on the scene, clean-shaven: the first larva with its incongruous mouth parts had survived. We became good friends, and saw each other again after he had settled in Amherst. We became colleagues once more, when we both taught at the embryology course at the Marine Biological Laboratory in Woods Hole in the 1940s.

Whereas the term *heteroplastic* designated transplantations between two species of the same genus, the term *xenoplastic* was introduced for exchanges between embryos that were more distantly related. As Spemann stated in the quotation above, these experiments make a novel contribution to our understanding of embryonic induction in that they bring into sharp focus the complex nature of this phenomenon. On the one hand, the inductor serves as a releaser: it activates *species-specific* potentialities inherent in the reacting system. On the other hand, the same experiments reveal an "instructive" aspect of induction: the ventral ectoderm responds in a *region-specific* way. The inductor in the mouth region directs differentiation to mouth structures. The same ectoderm will respond with the differentiation of lens, otocyst, or gill covering, if brought into contact with the appropriate local inductors. Obviously, the terms "instructive" and "releasing" are inadequate for an in-depth analysis of the phenomenon. Previous reflections on lens induction have made this point clear already. These conceptual difficulties could not be handled in 1921, with the material available at that time; and they have not been resolved to this day. This, however, does not detract from the significance of the results and the challenges they pose.

It is evident from the language used by Spemann in his discussions that he considered the hetero- and xenoplastic transplantation method primarily as a tool for the unravelling of the complexities of embryonic tissue interactions: this is, after all, the approach of the experimental embryologist. However, these experiments also have a bearing on another basic problem: the activation of genes during development. Spemann was aware of this fact, but it did not figure prominently in his thoughts. He might have been expected to address this issue in his speech before the annual meeting of the German Society for Genetics in 1924, entitled "Genetics and developmental mechanics." There is, indeed, a section on activation of the genetic material (1924a, p. 73). The thrust of the argument, however, is a repudiation of Weismann's theory of unequal nuclear division. He quotes the impressive array of his experiments that supports the alternative view that "the activation of the genetic material does not occur by autonomous segregation of hereditary units but by far-reaching interactions of parts, that is, epigenetically" (1924a, p. 78). Toward the end of the lecture, he suggested once more the exchange of mouth epidermis between anurans and urodeles, with the comment: "One can easily recognize that deep insights in the activation of hereditary factors can be gained in this way" (1924a, p. 78). That is all he has to say! When the experiment had become reality, the title of the joint publication with Schotté was "On xenoplastic transplantation as a means for the analysis of embryonic induction" (Spemann and Schotté, 1932).[1] The idea that the method could also be used as a means for analysis of gene activation was not mentioned in this paper, nor in his Silliman lectures (1936). Yet in his appraisal of the relations between experimental embryology and genetics, Spemann was more positive than most of his contemporaries. Outstanding embryologists, such as Frank Lillie, Charles M. Child, Ross Harrison, and Albert Dalcq, held the belief that the two disciplines had little in common. Indeed, one finds statements to the effect that Mendelian genetics and developmental mechanics are, in fact, incompatible. This is not the place to discuss the rationale on which the alienation between the two branches of biology in the 1920s and 1930s was based (see Hamburger, 1980a). It was left to a younger generation under the leadership of Ivan Schmalhausen, Richard Goldschmidt, Conrad Waddington, and others to establish developmental genetics as a legitimate and, indeed, central issue in developmental and evolutionary biology.

The experiment with Schotté remained Spemann's only involvement with xenoplastic transplantations. But his colleague and friend, Fritz Baltzer, took up the challenge. By extensive use of this method, he and his students made major contributions to what one might call evolutionary embryology. Baltzer, like Spemann, had been a student of Boveri's, and since his early work on hybridization in sea urchins had retained an interest in the role of the genome in development. Spemann and Baltzer became friends in Würzburg, and when Spemann moved to Freiburg in 1918, he invited Baltzer to join him and Baltzer accepted a professorship. In 1921 he returned to his native Switzerland as the Director of the Zoological Institute of the University of Berne. Very soon, he assumed a leading position in Swiss biology, and many of his students, foremost among them Ernst Hadorn, became prominent researchers. I was Baltzer's student in Freiburg, and took all his courses and seminars and emerged with a solid foundation in cytology

---

[1]The reader may wonder why Spemann put his name first, both here and in the publication of the organizer experiment. To him, the original idea deserved primacy over the execution of the experiment. This would apply to these two and to several other instances. On the other hand, in the case of O. Mangold and Spemann (1927), the authors had the idea of the experiment independently.

and genetics. Our friendship, which developed later, was sustained by regular correspondence and during his rare visits to St. Louis, my more frequent visits to his home in Bern, and vacations in the Swiss mountains, to which we both were very partial.

Baltzer had a wide range of research interests. He carried out work on xenoplastic transplantation between anurans and urodeles in the 1940s and 1950s in collaboration with a number of students. Briefly, the analysis of induction of mouth parts, balancers, suckers, and otocysts was carried to much greater depth than before, and was extended to the visceral skeleton, in which anurans and urodeles differ considerably. Two important generalizations were made: (1) induction systems show a great plasticity in evolution, and (2) evolutionary changes occur predominantly in the responses of the induced structures, whereas the inductors retain a more general and less specific induction capacity (review in Baltzer, 1952).

Spemann's publication of 1921 ends with a postscript, dated May 1921 and added in the proofs. It announces the first successful case of the organizer experiment performed by Hilde Proescholdt, in which the upper blastoporal lip of an early gastrula of the unpigmented *Triton cristatus* had been transplanted to the flank of an early *Triton taeniatus* gastrula. It had induced a perfect secondary neural plate in the pigmented host tissue, with an elongated white strip in its midline which extended to its posterior end. The embryo was then still alive; hence the contribution of the transplant to the mesodermal structures could not be ascertained. Spemann at that time introduced the term "organizer" in the following statement: "Such a piece of the organization center can be designated briefly as an 'organizer'; it creates an 'organization field' of a certain [axial] orientation and extent, in the indifferent material in which it is normally located or to which it is transplanted" (1921, p. 568).

The postscript raises a somewhat unsettling question: Why was the upper lip transplantation not performed along with the other heteroplastic transplantations in 1917? Why was it put off so long and eventually performed not by himself, but by Hilde Proescholdt, as the basis of her Ph.D. thesis?[1] The story of the discovery of the organizer would be incomplete, and deprived of an intriguing facet, if one were to sidestep this issue. I think one has to look for the answer in Spemann's deeply rooted misconception of the nature of the upper blastoporal lip of the early gastrula, which I have dealt with at length. It will be remembered that Spemann had erroneously identified the upper layer of the blastoporal lip of the early gastrula as neural ectoderm and as the differentiation center of the neural plate.

It is true that Spemann gave consideration to the alternative explanation: that the neural plate is induced by the subjacent mesoderm. And it is also true that the homeoplastic transplantation lacking a permanent marker did not permit a crucial test of this controversial issue. But now that heteroplastic transplantation provided a permanent marker, why was he content with the exchange between prospective anterior neural plate and prospective epidermis? Why did he stop short of repeating the transplantation of the upper blastoporal lip? To understand

[1]See the appendix.

this omission it may help to remember the chronology of events. The homeo-plastic transplantations were done in the spring of 1916 and the results published in 1918. According to a footnote, the heteroplastic experiments were done in 1917, his last year in Berlin-Dahlem. At that time he was still under the spell of his misconception, and it did not seem worthwhile to repeat the upper lip trans-plantations with the heteroplastic method. He needed no further proof for the establishment of the upper blastoporal lip as the organization center, and his strong commitment to the notion of neural plate determination entirely within the ectodermal layer mitigated his incentive to use the heteroplastic method for a test of the alternative hypothesis of neural induction by mesoderm.

The publication of the experiments of 1917 was delayed until 1921. (Since the microsurgery on early embryos was the domain of his laboratory, there was no danger then that priorities would be challenged. The situation changed soon there-after!) The delay was undoubtedly due to Spemann's move to Freiburg in 1919 and his preoccupation with many administrative duties. By the time the discus-sion of the experiments was written, he had undergone a gradual conversion; more evidence in favor of neural induction by the underlying mesoderm came to light, and the need for a test of this hypothesis by the heteroplastic method became urgent. But by then it was too late for him to remedy the omission. Other duties kept him away from the laboratory; there is no record of any experimental work of his own during the years 1918 to 1924.

Two events in particular prompted his change of mind. One was a letter by a colleague, the anatomist Hans Petersen of the University of Heidelberg, which was written after the appearance of the 1918 paper. In it, Petersen pointed out that "the superficial cells in the upper blastoporal lip would have been invagin-ated around the blastopore, and in all probability would have formed the roof of the archenteron, that is, notochord, somites, gut epithelium, *but not neural plate.*" (For Spemann's paraphrase of the letter see 1921, p. 550; italics mine. I have made exactly the same argument above.) The second piece of evidence came from a case of heteroplastic transplantation which, incidentally, was identical with one described in the homeoplastic series (Fig. 2–11). A round white *cristatus* implant, placed at some distance above the blastopore of a *taeniatus* gastrula, was trans-formed during neural plate formation into a long, narrow strip which extended to the posterior end of the neural plate. Spemann drew the obvious conclusion: "[during gastrulation] the posterior end of the implanted piece, which at the beginning of gastrulation had been at a considerable distance from the site of invagination at the upper lip, had moved close to the posterior end of the neural plate and the closed blastopore. Therefore the entire length in between must have turned inward and formed the archenteron roof" (1921, p. 548). This statement is almost a verbatim reiteration of a statement made in 1918 to explain an iden-tical case in the homeoplastic series. At that time, the message was not heeded, but now its relevance at last became clear.

> The entire transplanted piece [upper blastoporal lip] could belong to the inva-ginating region, hence represent presumptive endomesoderm; this would then be overgrown by host ectoderm . . . and it would have determined the neural plate in the covering ectoderm. The latter would therefore not be derived from

the transplanted piece but from the embryo to which the piece was transplanted. Therefore, if the piece [upper blastoporal lip] was taken from a *taeniatus* embryo and transplanted to a *cristatus* embryo, then notochord and somites should be formed by the former and the neural plate by the latter. This can be decided exactly. (1921, p. 551)

The postscript reports the first result of this experiment.

Briefly, then, the transplantation of the blastoporal lip had been done homeoplastically in 1916, but had not been repeated heteroplastically in 1917 because its true significance had not been perceived. The crucial importance of the experiment was recognized when the discussion was written in 1920 (the manuscript was submitted to *Roux' Archiv* in January, 1921), and the experiment was assigned to Hilde Proescholdt in the spring of 1921.

It is not possible very often to trace the history of an important discovery in such detail as I was able to do in this instance. Surely, each major discovery has its own unique history. But I wonder whether some ingredients, such as ingenious intuition and impeccable experimental design, side by side with errors of judgment, fateful omissions, and nonrational personal preferences, may not be shared by many. In spite of Spemann's singleness of purpose in pursuing his goal, the road to the discovery of the organizer was anything but smooth and straight, and there is a touch of irony in the thought that it was made at the sacrifice of one of his most cherished ideas.

CHAPTER 3

The Organizer Experiment (1924)

The discovery of the organizer, reported in the publication of Spemann and H. Mangold (1924),[1] was the fulfillment of Spemann's quest to comprehend the determination of the axial organs in vertebrate embryos that dates back to his earliest experiments. I have traced the steps that led gradually, and with admirable consistency, to the attainment of his goal, and I have tried to shed light on the circumstances which led to the assignment of the crucial experiment to his student, Hilde Proescholdt, in the spring of 1921.

Description of the Experiment

The transplantation of a piece of the upper blastoporal lip of the gastrula of the newt, *Triturus cristatus* (with unpigmented eggs), to the flank of a gastrula of the common newt, *Triturus taeniatus* (with pigmented eggs) (Fig. 3–1A,B), resulted in the formation of a secondary embryo initiated by the transplant and composed of a mixture of transplant and host tissues. The transplant which was a part of the organization center was designated as the organizer.

Altogether, five experimental cases were described in detail and a few others were mentioned in passing. In the most advanced and most complete case, designated *Um 132,* the donor was an advanced gastrula (this has to be remembered). The median part of the horseshoe-shaped blastopore was transplanted to the left flank of an embryo of the same stage, at some distance from the blastopore of the host. The transplant invaginated almost completely; a small number of cells remained on the surface. Two days after the operation, the host embryo had

[1]An English translation may be found in Willier and Oppenheimer, 1974.

48

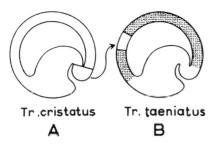

Tr .cristatus   Tr. taeniatus

A   **B**

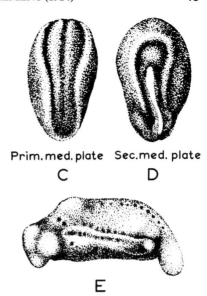

Prim. med. plate   Sec. med. plate

C   **D**

Fig. 3–1. The organizer experiment. A,B. Transplantation of the upper blastoporal lip of a gastrula of *Triturus cristatus* to the ventral side of a gastrula of *T. taeniatus*. C. Neural plate of host embryo. D. Secondary induced neural plate. A narrow strip of the unpigmented *cristatus* implant in the secondary neural plate. E. Secondary embryo *(Um 132)* on the flank of the primary embryo. From Spemann and H. Mangold, 1924; A and B added.

E

attained the neurula stage. A second plate with neural folds was clearly visible on the flank: it was pigmented except for a narrow strip of unpigmented tissue (Fig. 3–1C,D), that is, composed almost entirely of host tissue. A day later, both the primary and the secondary embryo had advanced to the tail bud stage. The axis of the secondary embryo ran nearly parallel to that of the primary embryo, but it was shorter than the latter. While the head of the primary embryo was normal, that of the secondary embryo was incomplete: it lacked the anterior brain parts and optic vesicles and ended at the level of the hindbrain, which was flanked by two otocysts (inner ear vesicles). Significantly, they were at exactly the same level as those of the host embryo. Both embryos had two clearly recognizable rows of somites and a tail bud. The embryo was sacrificed at this stage. The sections showed a complete set of secondary axial organs: neural tube, notochord, somites, pronephric tubules, and intestine. The latter was shared, in part, by the host embryo.

Two points are of special interest: on the one hand, the unitary organization of the secondary embryo, and on the other hand, the chimaeric structure of the secondary axial organs, that is, their composition of donor and host cells (Figs. 3–2, 3–3c). The neural tube was formed almost entirely of pigmented host tissue. The white strip was composed of transplant cells which had not invaginated. The notochord was unpigmented, that is, composed of transplant cells. Some of the somites were chimaeric, others were entirely formed by host cells or by transplant cells (Fig. 3–3c).

The term "organizer" is meant to give expression to these two aspects, with emphasis on the unitary organization of the secondary embryo. Spemann gives the following definition: "A piece of the upper blastoporal lip of an amphibian embryo undergoing gastrulation exerts an organizing effect on its environment in such a way that, if transplanted to an indifferent region of another embryo, it

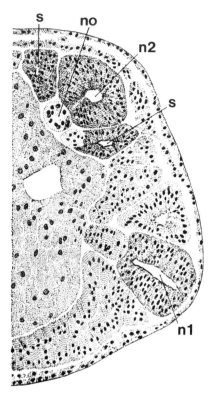

Fig. 3–2. Organizer experiment. Section through primary and secondary axial systems. n1, Primary neural tube; n2, secondary neural tube; no, secondary notochord; s, secondary somites. From Spemann and H. Mangold, 1924.

causes there the formation of a secondary embryonic anlage. Such a piece can therefore be designated as an organizer" (Spemann and H. Mangold, 1924, p. 637). The deficiency of the secondary embryo in the case described above can be regarded as a minor flaw. Very complete secondary embryos were obtained later by other investigators (Fig. 3–4). The absence of forebrain and eye in this case is probably due to the fact that the transplant was taken from an advanced gastrula; it did not include the anteriormost part of the mesoderm mantle which induces forebrain and eyes. Complete secondary embryos were obtained when the organizer was taken from an early gastrula.

The chimaeric nature of the secondary axial organs is also stressed in the discussion in the paper, but it is most clearly formulated in a brief report given in the same year at the meeting of the Association of German Natural Scientists and Physicians: "The secondary embryonic anlagen of the experiments of H. Mangold can be chimaeric in all their parts. The borderline between the induced *taeniatus* cells and the inducing *cristatus* cells can cut straight across a somite or the notochord. It looks as if such a secondary embryonic anlage has been built from the available material by a superior [übergeordnete] force without regard to its derivation or its species identity" (1924b, p. 1093).

Clearly, the organizer action is a very complex phenomenon in which cellular

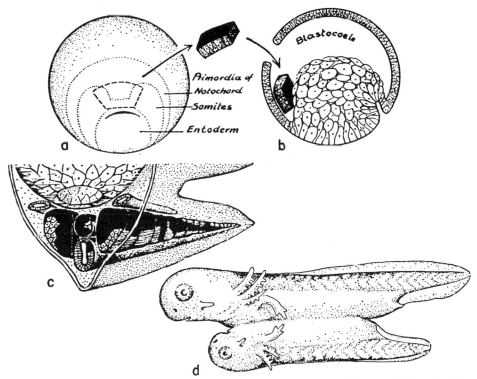

Fig. 3–3. Organizer implantation by the Einsteck-method. a,b. Upper blastoporal lip of early gastrula of *Triturus cristatus* implanted into blastocoele of *T. taeniatus.* c. Diagram of posterior part of secondary embryo. Note chimaeric somites. d. Primary and secondary induced embryo. From J. Holtfreter and Hamburger, 1955.

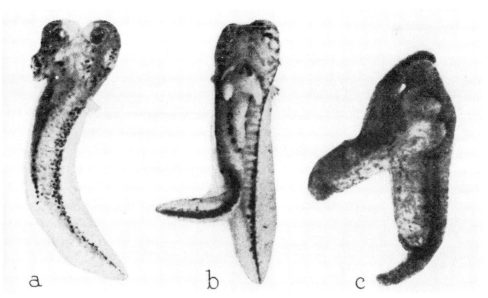

Fig. 3–4. a,b,c. Three examples of complete secondary embryos, resulting from transplantation of the upper blastoporal lip of early gastrula. From J. Holtfreter, 1933c.

and supracellular events will have to be given due consideration. But before considering how the organizer works, I shall discuss the other cases described in the organizer paper, to the extent that they reveal new facts. Unfortunately, the difficulty of raising operated embryos outside their jelly membranes, in an inadequate culture medium, limited the available material drastically. Moreover, the few other surviving embryos were sacrificed even earlier than the one described above, as a matter of precaution. In several instances, the organizer was implanted near the zone of invagination of the host embryo; as a result, the two anlagen were partly fused. This also complicated the interpretation. In two cases, the secondary neural plate showed a long narrow median strip of unpigmented *cristatus* material, which later formed the floor of the neural tube and was continuous with the mesodermal structures in the interior. The explanation is that the transplant was taken from an advanced gastrula. Its anterior end included prospective neural material which was not destined to invaginate, but merely to stretch and elongate. This it did in its new position. In two other cases, the secondary neural tubes were composed entirely of host cells; that is, they were induced by transplanted mesoderm. For this reason they are of particular significance (see Induction of the Neural Plate, below). The notochord was uniformly unpigmented, except for one case in which a few pigmented *taeniatus* cells were incorporated. The partly chimaeric composition of the somites was mentioned. This, then, was the altogether scanty material that Spemann had at his disposal. It is due to his ingenuity and his longstanding familiarity with the basic issues that he was able to draw the far-reaching conclusions that made the experiment famous.

Spemann's Analysis of the Experiment

Invagination

The capacity of the upper blastoporal lip for autonomous invagination had been observed before in the homeoplastic experiments of upper lip transplantation (1918). The gradual disappearance of the transplant was actually observed in some cases of the H. Mangold experiment. A chance observation gave Spemann the idea that the organizer might spread its influence on adjacent ectoderm cells immediately after its transplantation and induce them to participate in the gastrulation movements: "In one case, the secondary somites were much more darkly pigmented than the host somites; the thought suggests itself that they might have been formed by prospective ectoderm just like the secondary neural tube, which they resemble in color. One would then have to assume that the organizer had instigated the blastula cells of the animal [upper] hemisphere, where it had been implanted, to invaginate extensively and subsequently had determined their fate to become somites" (Spemann and H. Mangold, 1924, p. 627). In support of this notion, Spemann refers to an experiment by O. Mangold (1923) which gave direct evidence of the transformation of prospective ectoderm to mesoderm. Prospective ectoderm of the early gastrula was transplanted to the upper blastoporal lip of another embryo. The transplant participated in the invagination and later differentiated into mesodermal structures such as somites. Thus the transplant had changed its germ layer affiliation.

The incorporation of indifferent cells during the invagination of the organizer, if definitely proven, would have a significant theoretical implication. It would suggest that the organizer transplant, which represents only a part of the normally invaginating chordamesoderm area of the donor embryo, tends immediately to regain its original size by inducing adjacent ectoderm to participate in invagination. This would be a case of assimilative induction.

Self-differentiation

This term was introduced by Roux to denote the capacity of an embryonic structure to undergo differentiation according to its normal fate without the influence of extrinsic inductive agents. The term has been used widely by experimental embryologists; its definition seems to be straightforward, but the term is actually ambiguous in several respects. (1) It does not imply that the structure has had this capacity from the beginning of its development; it may have acquired it in a preceding stage by interaction with its environment. Hence it is important to specify the stage of development at which the term is applied. (2) An embryonic part, at a given stage of its development, can be self-differentiating with respect to a particular characteristic, but, later on, become dependent on an extraneous causal agent with respect to another aspect of its differentiation. For instance, Spemann considered the possibility that the blastoporal lip might acquire the capacity for mesoderm differentiation as a result of its preceding invagination: "The inducing capacity of the transplant [in the organizer experiment] could be exhausted with the impulse to invagination; everything else [mesoderm differentiation] would be solely the consequence of the secondary gastrulation" (Spemann and H. Mangold, 1924, p. 627). Later experiments showed that no such causal nexus exists. The important conclusion is that the organizer is self-differentiating with respect to both invagination and mesoderm differentation. (There seems to be no objection to applying the term "self-differentiation" to morphogenetic movements, such as invagination, since they are an integral part of the differentiation process.)

What would the organizer transplant have done, if left in place? Since it was taken from the median part of the upper lip, it would have formed notochord flanked by somites or parts of somites. In the H. Mangold experiments, the notochord was completely unpigmented save for one case in which a few host cells were interspersed. Evidently, this structure was truly self-differentiating. But the secondary somites gave a different picture: many of them were chimaeric, others were composed entirely of donor or entirely of host material. Only the donor-type somites can be said to have originated by self-differentiation; the chimaeric somites were partly the result of self-differentiation and partly of assimilative induction. The information on other mesodermal and on endodermal derivatives was too scanty to permit any conclusions. In summary, self-differentiation and induction processes interact in a complex fashion to create the secondary embryos.

Spemann detected another feature in the induced embryos which indicated that the transplant material behaved differently from what one would have expected:

> *A priori,* there exists the possibility that the transplanted piece develops by pure self-differentiation to exactly the same parts which it would have formed at the place from which it was taken, and that it would have added from the indifferent neighborhood that which it lacked to make a whole [secondary embryo]. Such a perfect self-differentiation of the organizer, however, does not take place; otherwise the transplant would later be too large for the smaller secondary anlage. Since the transplant fits in harmoniously, its material has been disposed of differently from its normal development. (Spemann and H. Mangold, 1924, p. 632)

In summary: while self-differentiation remains a significant component in the formation of the secondary embryo, it is intricately interwoven with both inductive and regulatory processes. The integration of these three components in the formation of a harmonious whole goes to the heart of the organizer problem.

Assimilative induction

This term does not appear in the discussion of the organizer paper, but it reflects accurately what Spemann had in mind as the explanation of the origin of chimaeric somites in secondary embryos. It means simply "determinative effects progress from cell to cell" (Spemann and H. Mangold, 1924, p. 633). The term is then used in the chapter on the organizer in his book (1938, p. 163). By a strange inconsistency, Spemann used the term "assimilative induction" in another part of his book with an entirely different meaning. He and O. Mangold had discovered that a piece of neural plate, when implanted in a gastrula, will induce neural plate in ectoderm, by contact. They called this "homeogenetic or assimilative induction" (Mangold and Spemann, 1927; see also Spemann, 1938, p. 214). I shall use the term in its original meaning, that is, synonymously with "appositional growth."

The concept of assimilative induction has always played a major role in Spemann's thinking; it dates back to his idea of "appositional growth," which was conceived early in his experimental work. I have dwelled at length on his tenacious but—as it turned out—misguided endeavor to explain neural plate induction in these terms. It is one of the ironic twists which happen occasionally in science, that the organizer experiment exposed the fallacy of this explanation of neural plate determination, and at the same time provided an excellent example of assimilative induction in the mesoderm, and thus, at last, a vindication of his cherished early intuition. Indeed, there is only one explanation for the origin of chimaeric somites and of the somites composed of host cells: it has to be assumed that the invaginated transplant mesoderm, being incomplete (since it represented only part of the chordamesoderm of the donor embryo) recruited adjacent host mesoderm to produce a complete set of two rows of somites in the secondary embryo. But what exactly was the origin of the "assimilated" mesoderm? It was carved out—to use Spemann's term—from lateral plate mesoderm of the host embryo. (The lateral plate is the lateral and ventral mesoderm sheet which in normal development gives rise to the urinary system, the walls of the coelom, and the heart.) Spemann reminds us also in his discussion that he had suspected in one case the incorporation of prospective ectoderm cells of the host in an induced somite and that O. Mangold (1923) had shown that prospective epidermis, when transplanted to the upper blastoporal lip, participates in the invagination and in the formation of mesodermal structures, obviously under the influence of its new environment. Hence assimilative induction is by no means limited to the mesoderm.

Induction of the neural plate

Spemann's dilemma was to decide whether neural plate determination results from assimilative induction or from an inductive action on the part of the subjacent mesoderm. The problem was aggravated by his undisguised preference for the first alternative, whereas the experimental data tended more and more to favor the latter. One might have expected that the occurrence of neural plates composed entirely of host material would have removed all doubts, but this was not the case. It has puzzled me that in the discussion in the Spemann–H. Mangold paper the theme of neural plate determination is underplayed, and a considerable degree of ambivalence still prevails. This is apparent in the following quotation:

> The ectodermal constituent [of the transplant] could have developed by self-differentiation and formed the narrow strip, which in turn could have caused the ectoderm situated anterior and lateral to it to form progressively neural substance. However, the determination could have started in the subjacent endo-mesoderm and affected the *cristatus* and *taeniatus* components of the overlying ectoderm alike. And finally, it is conceivable that the substratum is necessary for the first [step in the] determination and that it spreads from then on purely in the ectoderm. (Spemann and H. Mangold, 1924, p. 606)

The reference at the beginning is to two cases mentioned above, in which the otherwise pigmented, that is, host-derived, neural plate showed a narrow median strip of unpigmented transplant tissue. Obviously, the problem remains unresolved in his mind, and, to complicate matters further, a third alternative is added which is actually a combination of the other two. But what happened to the two critical cases in which the neural plate consisted entirely of host cells? They were almost completely forgotten! I found only a brief reference to them in the section on the structure of the secondary embryo: "In the neural tube, the tissue of the host prevails; *cristatus* cells can be absent completely (two cases) or they form only a small strip" (Spemann and H. Mangold, 1924, p. 626). The obvious step, to conclude from the two cases that the neural plate must have been induced by the underlying mesoderm, was not taken.

The matter took a decisive turn when the results of an entirely different type of experiment became available. In pursuit of another problem, Spemann and O. Mangold had devised an implantation method which became known as the "Einsteck-method." A transplant was slipped into the blastocoele (the inner cavity) of a blastula or early gastrula through a small slit in its upper hemisphere (Fig. 3–3a,b). The opening heals immediately, and the gastrulation movements of the host embryo result in the gradual replacement of the blastocoele by the invaginating and expanding archenteron (primitive gut). In this process, the transplant becomes attached to the inner surface of the ventrolateral ectoderm. If it has the capacity for neural induction, it will induce a secondary neural plate in this region. The advantages of this method are obvious: it is technically much easier than transplantation, and it can be used to test a variety of tissues which cannot be transplanted to the surface of a gastrula, such as adult tissues. In fact, the method became an indispensable tool in the following decade. Who was the originator of this method? Both Spemann and O. Mangold claimed priority in different publications. But a remark of Spemann sounds very credible: "The method which has become very important has been used for years in our institute. Whether the first idea came from O. Mangold or from me cannot be ascertained by us any more; perhaps it was found in conversations" (Spemann and Geinitz, 1927, p. 131).

At the suggestion of Spemann, his colleague Bruno Geinitz used this method to implant the organizer into the blastocoele. The transplant induced secondary neural plates and secondary embryos as complete as those obtained by H. Mangold. Geinitz went a step further: he obtained the same result when he used organizers from frog and toad donors and salamander gastrulae as hosts. Incidentally, it was on this occasion that the term "xenoplastic" was coined by Geinitz to designate interchange of transplants between species which were more distantly related than species belonging to the same genus (Geinitz, 1925).

While this experiment gave rather convincing evidence for neural induction by contact rather than by assimilative induction, any further doubt seemed to be removed when Spemann's student Alfred Marx used unmistakable mesoderm as the neural inductor. He removed the dorsal ectoderm in a *late* gastrula and thus exposed the underlying mesoderm mantle. He then cut out small pieces of the mesoderm and slipped them into the blastocoele of an early gastrula. The experiment was done heteroplastically. A neural plate was induced in host ectoderm in every case; control experiments with ectoderm transplants were negative (Marx, 1925). Now Spemann seemed to be convinced. "If a piece of endomesoderm can exert this effect on the overlying ectoderm, it is probable that . . . in the normal development, likewise, the neural plate is determined by the subjacent endomesoderm" (1924b, p. 1092).

O. Mangold had been one of Spemann's first students at the University of Rostock before the First World War. He had served in the Air Force throughout the war and had joined Spemann again in Freiburg. The former student now became a colleague and a close personal friend. And they were linked by a strong professional bond: the common interest in the complexities of early amphibian development and the challenges of the experimental analytical approach opened up by Spemann. The two were very different personalities, and they came from very different backgrounds. Perhaps this fostered, rather than impeded, their mutual sympathies. Besides, they were both Swabians, and their birthplaces were not far apart. Ethnic closeness—of the Prussians in the north and the Swabians, Franconians, and Bavarians in the south—has always played a role in personal social relations in Germany. Spemann was tall and slender. Although he had fully recovered from an early bout with tuberculosis, he was never of robust health and he had to husband his strength. For instance, he avoided working in the evenings. He came from an affluent family; his father owned a well-known publishing house in Stuttgart, a fairly large city. He had grown up in a cultivated home frequented by artists and writers. Later on, he created his own home in the same style. I remember life-size panels of reproductions of Dürer's "Four Apostles" in the entrance hall of his house in Freiburg, a large bust of the poet Schiller by a well-known artist, and a handsome library of leather-bound volumes. He had started out in his father's business, but soon discovered his vocation for the natural sciences. For practical reasons he had begun to study medicine, but he had soon switched to zoology and pure science.

O. Mangold, on the other hand, was sturdy and vigorous. He came from peasant stock and had grown up on a farm. He was forceful in everything he did, from experimenting to skiing. He did not have Spemann's flair and originality, and he did not break new ground. But he had a keen mind and a thorough comprehension of experimental embryology, and he was an excellent experimentalist. He made some important contributions, one of which was mentioned above. Around 1930, he wrote several extensive review articles which were among the first comprehensive accounts of the state of the art.

Mangold was appointed as the director of the Division of Experimental Embryology at the Kaiser Wilhelm Institute in Berlin-Dahlem in 1924. He had married Hilde Proescholdt in October of 1921, and they left Freiburg early in 1924. When I worked in Mangold's department in 1926 and 1927, I had the title of assistant, but I was independent in my work. Mangold gave me unrestricted freedom and generous support. We pursued our different lines of research; I continued my experiments on the development of the nervous system and he worked on embryonic induction. I do not remember that we interacted extensively. We respected each other, but we were personally not very close. In our case, our different backgrounds and personal interests were not conducive to a close friendship. Mangold was reserved by nature and, still mourning his wife's death, rather withdrawn. I took advantage of the vibrant cultural life of Berlin, which was then a cosmopolitan center of the arts: of expressionism in painting and literature, Max Reinhardt's stage innovations, and Mary Wigman's modern dance choreographies.

Mangold became Spemann's successor twice (in Berlin-Dahlem and in Freiburg), and he was also his biographer (O. Mangold, 1953). After he had left for Dahlem in 1924, he remained in close contact with his mentor. Spemann was instrumental in the appointment of Mangold as his successor in Freiburg in 1937, which is evidence of his unqualified confidence in Mangold's scientific and administrative abilities. At that time, the Nazi regime was firmly in power. Mangold supported the Nazi ideology and soon became active politically at the university. In 1938 he was elected by the faculty to the office of rector of the university, a honorific administrative position which, like all other institutions, had become politicized. He held it until 1941, apparently to the satisfaction of the party and the faculty, since the term of the rectorship was traditionally only one year. Thereafter, he continued as director of the Zoological Institute and resumed his teaching and research. After the end of the war, his identification with the Nazi regime cost him his job. He was dismissed from the university by the French, who occupied South Germany in the spring of 1945. He spent the rest of his life in a small, privately financed research institute in Heiligenberg, near Lake Constance. He and his asso-

ciate, Carl von Wöllwarth, published extensively, mostly on embryonic induction. He died in 1962.

When I returned to Freiburg at the end of 1927, as a faculty member with the rank of assistant (professor), the complexion of the Zoological Institute had changed considerably. The senior staff members were Fritz Süffert and Bruno Geinitz. Both were in their forties. In addition, there were several foreign guest workers: Oscar Schotté from Geneva, who then moved to the United States; Fritz Lehmann from Bern; and Tadao Sato from Japan, who worked on lens regeneration. Furthermore, there were several Ph.D. candidates. We were a closely knit group. Soon a special friendship developed between myself and the Süfferts.

Süffert's interest was in adaptive coloration, particularly of butterflies and moths, and he became a leader in this field, both as an experimentalist and as a theorist. Next to Spemann, he was the most stimulating and original member of the group. I remember many animated discussions with him and Spemann during the daily afternoon tea hours in Spemann's reprint room. One of the recurrent themes was the evolution of adaptive traits. Spemann was never convinced that natural selection could adequately explain very complex adaptations such as mimicry. He listened patiently to Süffert's arguments in its favor and he made his own counter-arguments; but, as was to be expected, neither one convinced the other. Süffert and his wife were both artistic: he was an excellent photographer, and she did the many beautiful illustrations in color of caterpillars and pupae for his publications. She also did wonders in rearing the offspring of rare butterflies and moths through molting and pupation to the final metamorphosis to the adult stage. But the most difficult feat was to persuade them to breed in captivity. She did this and thought that her success was due to presenting them with the scent of fresh flowers in the cages.

The Süfferts came from Günsbach, a small town near Strasbourg in Alsace. They had befriended a fellow countryman, Albert Schweitzer, the famous Christian exegete, philosopher, writer, biographer of Johann Sebastian Bach, organist and organ builder, and eventually the leader of a hospital group in the West African jungle, in Lambarene, after he had obtained his M.D. in middle age. He was the recipient of the Nobel Peace Prize in 1952. If I remember correctly, the Süfferts had been married by him. Süffert met a tragic fate. He had moved to Berlin as a member of Goldschmidt's Genetics Department at the Kaiser Wilhelm Institute for Biology, and later became an editor of Naturwissenschaften. *During the last weeks of the war he and thousands of other middle-aged men were drafted and, poorly armed, lost their lives in the senseless effort to defend Berlin against the tide of the invading Russian army—one of the last crimes of the Nazi regime.*

Bruno Geinitz was an entomologist by vocation and a specialist in honeybees and the practical aspects of honey production. I suspect that his brief interlude in experimental embryology was Spemann's doing. He probably enlisted Geinitz in the middle 1920s when there were not enough hands to bring in the harvest from the organizer experiment. As we saw, the draftee proved to be up to the mark. But his heart was with the bees. Later on, he founded an Institute for Honeybee Research which was connected with the Zoological Institute. His research turned out to be very valuable and profitable for the Black Forest beekeeping industry. And there was never a shortage of delicious honey at the daily tea hour in the reprint room.

The Geinitz house was close to that of the Spemanns on the hillside of the Lorettoberg in a suburb, with a free view of the mountains and the city and with the Münster below. The friendship of the families deepened after the Mangolds had left. Geinitz contributed a great deal to the good spirit in the Zoological Institute. There were a few strange fellows among the graduate students, but altogether, the atmosphere was very harmonious. Since the group was small and congenial, the excitement of the new discoveries was shared by us all with genuine enthusiasm. The only sufferers were the innumerable amphibian embryos on which we practiced our microsurgery, most of which died in vain.

Regulation

This concept has held a central position in experimental embryology from its very beginnings. I have referred to its role in motivating Spemann's earliest experiments. Regulation denotes the capacity of fragments of an embryo to restore the whole. If applied to blastomeres or parts of somewhat older embryos, it implies the restoration of the whole embryo. We speak also of regulation following the ablation of part of an organ primordium, such as the limb anlage or the anterior neural plate. It should be understood that the term regulation designates a phenomenon that is a particular aspect of the epigenetic nature of development, but that it has no analytical or explanatory significance. It is probably for this reason that it is not used by Spemann in the analysis of the H. Mangold experiment. In a later publication, however, he clearly recognizes the role of regulation: "If a whole embryonic anlage with eyes and otic vesicles can be induced by a small piece of the upper blastoporal lip, on the ventral side of the host embryo, then there is undoubtedly a participation of regulatory forces which tend toward wholeness" (1931a, p. 446).

We find, however, some more specific references to regulation in which the term itself is not used. One instance has been referred to already: since the secondary embryos are smaller than the primary embryos, size regulation must have occurred. Furthermore, one could consider the formation of the chimaeric mesodermal axial organs in secondary embryos as an example of regulation, but again this would not help us to understand the process any better than by describing it as a combination of self-differentiation and assimilative induction. The situation changed, however, when a modification of the original organizer experiment showed convincingly that, indeed, regulative capacity is an inherent attribute of the organizer. The experiment was done by Hermann Bautzmann. He was an assistant in Spemann's institute and a friend of mine. He found that the lateral parts of the upper blastoporal lip are as capable as the median strip of inducing a secondary embryo. In fact, the extent of the area that has this capacity coincides with the extent of the chordamesoderm area as outlined in Vogt's fate map. If a lateral part of the upper blastoporal lip is to become the central axis of a secondary embryo, then it is obvious that regulation has to occur and that the lateral piece has to acquire bilateral symmetry. Instead of forming notochord at its inner margin and one row of somites at its outer margin, it now forms notochord in the center, flanked by two rows of somites (Bautzmann, 1926). In the reference to this experiment in his book, Spemann speaks of "a *regulation* in which either the inducing fragment [of the upper lip] transforms itself into a whole or complements itself by including adjacent tissue" (1938, p. 146).

Structure

Does the organizer have a structure? It is not immediately obvious what Spemann had in mind when he raised this question; we have to look for the criteria he used. In the discussion of the organizer paper, the question of structure was linked to the orientation of the induced (secondary) embryo with respect to the main axis of the host (primary) embryo:

> The orientation [of the secondary embryo] could be caused entirely by the host embryo, entirely by the transplant, or by both. . . . If the first alternative should hold true, then the transplant would be structureless and behave entirely passively. . . . According to the second and third alternatives, the implanted organizer would possess a distinct structure, on which would depend the direction of its invagination and longitudinal stretching. (Spemann and H. Mangold, 1924, pp. 628–629)

In other words, the capacity for autonomous directional invagination and stretching was taken as evidence for a polarized structure of the organizer. In his address before the International Congress of Zoologists in Budapest in 1927, Spemann was more explicit: longitudinal structure meant anterior-posterior polarity, perhaps in the form of a gradient. The observation that the axis of the secondary embryo could be at an oblique or even a right angle to that of the primary embryo constituted the major argument in favor of an intrinsic polarized structure of the organizer.

But the cases in which the axes of the primary and secondary embryos ran parallel to each other suggested an interference of the former. The host influence could be considered merely as a mechanical deviation of the direction of transplant invagination by the more massive (invaginating) host mesoderm, and thus a trivial matter. But the embryo *Um 132,* in which the two pairs of otocysts of the primary and secondary embryo were exactly at the same level, had alerted Spemann to the possibility that the host embryo may play a more active role in the axial orientation of the secondary embryo. He found this problem so interesting that he devoted an extensive investigation to it (1931a).

In addition to axial structure, Spemann attributed laterality to the organizer. This claim was based on an experiment of Kurt Goerttler, a collaborator of Vogt. He replaced a lateral half of the upper blastoporal lip, for instance, the left one, with a lateral half from the other side of the lip, that is, a right half, taken from another embryo. In other words, he constructed blastoporal lips composed of two adjacent right or left halves. This resulted in the induction of two parallel left or right neural folds. Since the anterior neural folds are curved inward, their laterality could be easily identified (Goerttler, 1927). Spemann concluded, "Hence, the left half of the upper blastoporal lip must have had an intrinsic structure which endowed it with leftness; it must possess laterality" (1927, p. 948).

Spemann's concern with the structure of the organizer may seem antiquated today; but it was certainly a live issue at the time, and it had other far-reaching consequences. Carrying his reasoning a step further, he raised the question of whether retention of the structural integrity of the cells is a necessary prerequisite for organizer activity. This idea inspired the crucial experiment of destroying the structure of the organizer, which will be dealt with below.

Regional determination of the neural plate

In addition to polarity and laterality, Spemann postulated a regional structure of the organizer, "whereby the different regions of the mesoderm, as well as those of the neural plate induced by it, would be defined" (1927, p. 948). This idea occurred to him when he observed that the oldest embryo of the H. Mangold experiment *(Um 132)* was incomplete in that the anterior head in front of the ear vesicles was missing, whereas in a later organizer experiment of his own he obtained a secondary head that ended blindly behind the otocysts. "These different results could be due to differences of the inducing organizers that were taken from different regions of the organization center. That material of the mesoderm which later on comes to lie beneath the head region of the neural plate could be a *'head organizer'* whereas the material [that comes to lie] beneath the trunk could be called *'trunk organizer'*" (1927, p. 948). Spemann proceeded immediately to test this hypothesis. I include this experiment under the heading of "Spemann's analysis of the organizer experiment" because it is very closely related to it, although it was done several years later, in 1927 and 1928, and the results were not published until 1931 (1931a).

In one set of experiments, Spemann used the upper lip of the *early* gastrula, which would come to lie in the head region, and in another set he used the upper lip of the *late* gastrula, which would normally underlie spinal cord. The experimental design was complicated by his intent to test at the same time a possible effect of the site of implantation on the outcome. The upper lip of the early or the late gastrula, respectively, was implanted either to the head level or to the trunk level of the host embryo. In one set, the transplants were slipped into the blastocoele; this brought them also in contact with ventral ectoderm. The experiments revealed a distinct tendency of the prospective anterior mesoderm to induce head and brain structures, and of the prospective posterior mesoderm to induce spinal cord and tail structures. These tendencies, however, were labile: they could be overriden by local conditions prevailing at different host levels. For instance, prospective trunk mesoderm was capable of inducing brain structures at head levels and spinal cord at trunk levels.

Spemann's results were confirmed and extended by O. Mangold (1933). He avoided some of the complications of Spemann's experiments by using the Einsteck-method of implantation in the blastocoele, and by taking the transplants from the completely invaginated mesoderm of the neurula. He removed the ectoderm and divided the mesoderm mantle into four quarters along the rostrocaudal axis (Fig. 9–1). The anteriormost quarter, consisting of endoderm and the so-called prechordal plate mesoderm, had little inductive capacity. The second quarter induced preferentially brain structures and sense organs, including eyes, nose, otocysts, and, in addition, balancers. The third quarter induced mostly hindbrain with otocysts, and also anterior spinal cord. The fourth quarter induced spinal cords and tails. Thus regional induction specificity was clearly demonstrated. There was, however, a considerable overlap; for instance, eyes appeared in several cases of third-quarter implants and otocysts in second- and third-quarter transplants.

The follow-up of the organizer experiment had thus revealed that (a) the different regions of the chordamesoderm have, indeed, region-specific inductive capacities; and (b) these capacities go beyond the induction of neural structures. Anterior mesoderm induces not merely brain and sense organs, but also balancers in salamander embryos and suckers in frog embryos. Likewise, induction by the posterior mesoderm was not confined to spinal cord, but it included the trunk and tail with fins. Thus Spemann's original designation of "head" and "trunk" (better: trunk-tail) organizer was very appropriate.

Later experiments have shown that the chordamesoderm is "instructive" only in a limited sense. What is induced is not a detailed pattern, such as forebrain and eyes, but broadly delineated regions which have the characteristics of equipotential systems or morphogenetic fields. By definition, fields are regulative within their confines but self-differentiating as a whole. In this sense they do not require extrinsic agencies for their further differentiation. Rather, they undergo a process of stepwise segregation into smaller and smaller subunits until all their derivatives are irreversibly programmed. This process is called "self-organization," but, beyond naming it, it is poorly understood to this day.

One can go a step further and ask whether the role of the mesoderm in neural induction is confined to the labile determination of a set of regional fields, or whether it is involved in the subsequent phase of segregation by some nonspecific activity. Considerable information on this point is available for the rostral part of the neural plate, which can be designated as the "forebrain-eye field". As early as 1923, Spemann had an inkling of the field character of this region. "Much can be said in favor of the assumption that the eye anlage in the neural plate is a harmonious-equipotential partial system, because one obtains a whole eye, even if one transplants only part of the eye anlage" (1923, p. 13).

We owe an in-depth analysis of this particular topic to Howard Adelmann. Before he started a distinguished career as a historian of biology (he is best known for his studies and translations of embryological treatises by Fabricius and Malpighi), he conducted a series of experiments on eye determination which unfortunately are now largely forgotten. They were begun in Spemann's laboratory when he was a guest in the late 1920s. However, his experiments were not inspired by Spemann; they were the continuation of an earlier study of the morphogenesis of the vertebrate head which he had undertaken in collaboration with his mentor, B. F. Kingsbury (Kingsbury and Adelmann, 1924). The experiments were done on the American salamander, *Ambystoma punctatum*. They were based on the then available fate maps of the eye-forming areas in the neurula stage. They are located in the rostral part of the neural plate with rising folds, at some distance from the median plane. Adelmann divided the entire width of the rostral band into three equal parts (Fig. 3–5). The median strip represents mainly prospective

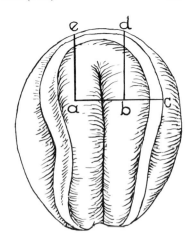

Fig. 3–5. Subdivision of the anterior part of the
neural plate into three parts, in the experiments of
Adelmann (see text). From Hamburger, 1960.

forebrain and small sections of the eye area. The lateral strips represent large sections of the eye-forming region at their inner border, and otherwise prospective
forebrain.

Adelmann was interested mainly in eye determination. To test the eye-forming
potential of the different regions of the anterior neural plate, he transplanted
median or lateral strips to the flank of another embryo. In one set he transplanted
the neural layer alone, and in another set together with its mesodermal substrate.
The transplants of pure neural ectoderm confirmed the field character and the
regulative capacity of the forebrain-eye region: both median and lateral strips
would give rise to eyes and forebrain. But to his surprise he found that the median
strips, which normally contribute little to eye formation, produced a high percentage of eyes (70%), whereas the lateral strips, which normally contribute substantially to eye formation, produced a low percentage of eyes (11%). Hence the
eye-forming potentiality seemed to be distributed as a gradient, with its peak at
the midline (Adelmann, 1929). For a better understanding of what follows, it
should be mentioned that all transplants of the median strip produced only single
eyes, attached to a forebrain vesicle. These strange findings raise an obvious question: How does the eye formation become centered in two bilaterally symmetrical
spots at some distance from the midline in normal development? Is this part of
the "self-organization" process of the forebrain-eye field, or is the underlying
mesoderm involved? The first hint that the latter is the case came from the experiment in which the lateral strip was transplanted together with its substrate. The
frequency of eye formation increased fivefold, from 11% to 54%. This means that
the anterior mesoderm is capable of reinforcing eye differentiation at a particular
spot of the forebrain-eye field. This combination, however, again produced only
single eyes. The decisive clue to an understanding of the origin of bilateral eyes
came when median strips were transplanted together with their mesodermal substrate. These pieces, which without substrate had produced only single eyes
(though with high frequency), now gave rise to two eyes in the great majority of

cases. They were connected with a forebrain vesicle, as in the normal head (Adelman, 1930). Since the wound in the donor embryo usually healed over and a normal head developed, donors and transplants together could give rise to four normal eyes—an impressive demonstration of regulation within a field.

One arrives at the conclusion that the same rostral mesoderm that is responsible for the initial induction of the forebrain-eye field continues to exert its influence on the neural plate; it focuses the diffusely distributed eye-forming potency on two areas, which thus become the two eye primordia. Adelmann (1932) made a careful study of the development of the prechordal mesoderm. He observed that two mesoderm condensations are formed underneath the prospective eye regions. It is very likely that these condensations are responsible for the "focusing" of the eye-forming potency on the two regions which actually form the eyes. It would be misleading to designate this effect as an induction. We are dealing with an auxiliary determinant which stabilizes the preexisting eye-forming potencies at their final location. Nevertheless, this secondary effect is of signal importance. If it fails, the unfortunate result is a single median (cyclopean) eye (Adelmann, 1936).

To make my point, I have been highly selective in choosing from a voluminous literature on eye determination. Some investigations support points made by Adelmann (for instance, O. Mangold, 1931; Alderman, 1935), but none equals his achievement.

It is of interest that a secondary influence of the mesodermal substrate on morphogenesis in the neural anlage has been demonstrated also for the trunk region. In normal development, the early neural tube, whose wall is at first of uniform diameter, is transformed into the spinal cord by widening of the lateral walls, which alone produce neurons and glia, and thinning of the ventral and dorsal ependymal layer to form what are known as the floor plate and roof plate, respectively. This transformation occurs only if a normal notochord is present; if its formation is suppressed or if it is excised, then the tube retains its uniform diameter. If it is underlain by muscle tissue, it develops an abnormal ventral thickening (Lehmann, 1926, 1928; J. Holtfreter, 1934a).

Spemann's Theoretical Evaluation of the Experiment

An organizer theory?

The most rewarding aspect of the organizer experiment was its potential as a catalyst. It raised new questions, it generated new ideas, and, above all, it suggested a multitude of new experiments. After 1924 the activities in Freiburg went into high gear, and an increasing number of graduate students and guests from abroad joined the old guard. The collector of the local newts, a laboratory employee by the name of Hund, was kept extremely busy. (One American visitor, noting that "Hund" is the German word for dog, thought that "Mr. Dog" was the generic name we used for our animal keepers). Of course, the female newts were returned to the ponds after they had laid their eggs in our aquaria, each egg folded neatly in a leaf of a water plant, *Elodea;* many were then recaptured after a few weeks. Herr Hund had a rotation schedule for this enterprise.

While the experimental work flourished, Spemann was not inclined to rush into far-reaching theoretical speculations. His restraint was understandable and justified. The less than a dozen experimental cases of the H. Mangold experiment did not provide sufficient factual information; it was much too early for broad generalizations. "The causal relations in the formation of the secondary embryos are still completely in the dark. The only thing that is certain is that somehow an induction by the transplant had occurred" (Spemann and H. Mangold, 1924, p. 627). But his reluctance to theorize had deeper roots, and this is perhaps the place to comment on them. The greatest strength of Spemann was his analytical acumen. In every experiment, the implications of the data were explored exhaustively; all possible interpretations were considered and weighed for their validity, and new experiments were planned to decide between alternative explanations. If I had to name the most important lesson I learned from him, it would be the thorough exploitation of experimental data. But with all attention to detail, the broad conceptual view was not lost sight of. He did not hesitate to formulate new concepts and to coin new terms. To this extent, analysis and synthesis went hand in hand. What he eschewed was extending his discoveries to general theories of development. Even in his book of 1936, (English version published in 1938), when a great deal of new information had been gathered, he resisted this temptation: "I wish to emphasize that as far as I know I have never drawn up an 'organizer theory.' I have coined the term in order to designate certain new and very remarkable phenomena which I have encountered in my experiments. From the beginning, however, I have considered the concept as preliminary. I distance myself from all attempts to make it now the building block of a theory" (1938, p. 367). Interestingly enough, the German title of his book *Experimentelle Beiträge zu einer Theorie der Entwicklung (Experimental Contributions to a Theory of Development)* seems to indicate that he considered the idea of a general theory of development as not entirely utopian. I wonder whether the change of the title in the English edition to *Embryonic Development and Induction* was prompted by greater caution or merely by the expediency of a shorter title.

Breach in the harmonious-equipotential system

The choice of the term "organizer" is often thought to have vitalistic undertones. While there is little doubt about Spemann's personal psychovitalistic creed (Spemann, 1943, p. 167; Hamburger, 1969), the question of whether his private philosophy influenced his scientific thinking requires careful scrutiny. The last sentence of the organizer paper raises this question. In order to retain the context, I quote the entire last paragraph:

> For the moment, it is of secondary importance whether the concept of organizer and organization center will still prove to be appropriate with advancing analysis, or whether they will have to be replaced by other designations which go into more detail. One can state already, however, that the concept of the organizer is the fundamental one, and that the term 'organization center' is merely supposed to designate the embryonic area in which the organizers are assembled at a certain stage, but not the center from which development is being directed. The designation 'organizer' (rather than perhaps 'determiner') is supposed to

give expression to the idea that the effect emanating from these preferential regions is not only determinative in a definitive, restricted direction, but that it possesses all those enigmatic characteristics which are known to us only from organic nature. (Spemann and H. Mangold, 1924, pp. 636–37)

Does the last sentence imply that in his opinion the components of organizer action that have been listed (self-differentiation, induction, regulation) do not suffice to explain the holistic aspect of organizer activity? Does he consider an additional, perhaps a vitalistic, principle?

When Driesch was confronted with exactly the same problem of regulation, he had opted for a vitalistic solution. On the basis of his discovery of the extreme regulative capacity of sea urchin and other embryos, he had formulated a theory of embryonic development which is epitomized in the concept of the *harmonious-equipotential system*. Equipotential means that each part has equal capacity or potentiality to substitute for any other part, and harmonious means that the end result of the interplay of parts, or of total regulation of a part, is the formation of a harmonious whole. Driesch argued that if such a system, as for instance a blastomere, is structurally homogeneous, if each part is identical with any other, then the system would be devoid of a structural or material cause for change or differentiation. Hence he invoked a nonmaterial causal agent which would set differentiation in motion. He called it *Entelechy* (meaning that which carries the end in itself). Spemann, as well as Harrison (1918) adopted the term harmonious-equipotential system. ("The organizer aims at wholeness, hence it behaves as a harmonious equipotential system, in the sense of Driesch" (1927, p. 948). However, they, along with most other experimental embryologists, rejected the vitalistic interpretation of the experiments. In his book Spemann states:

> Driesch felt compelled to conceive of the agency which produces the whole as non-spatial, as an 'intensive manifoldness' for which he reintroduces the old term 'Entelechy'. This is basically not much different from what was meant by the (likewise non-spatial) *idea*. Even if one does not believe that one must draw this ultimate consequence with Driesch, the equipotential system which is capable of harmonious patterning, remains a real problem which cannot be disposed of by the fact that the equipotential systems are perhaps never quite as harmonious nor quite as equipotential as Driesch thought them to be. (1938, p. 347)

The last line gives a hint of what Spemann considered as the critical fallacy in Driesch's reasoning: in the abstract, harmonious-equipotential systems may be structurally (or biochemically) homogeneous; but real systems could well be endowed with a preferential region that would initiate differentiation. The notion of a preferential region was introduced by Boveri as early as 1901 to explain polarity in Driesch's own material, the sea urchin embryo (see Spemann, 1938, p. 142). The organization center would be the equivalent of a preferential region in the amphibian gastrula, which is also a harmonious-equipotential system, inasmuch as its constriction in the median plane results in the formation of two whole embryos. Thus Spemann went beyond the mere rejection of the vitalistic interpretation; he took a decisive step towards a mechanistic explanation of the harmonious-equipotential system. In the organizer paper, he mentions only in passing his hope that the organizer experiment "might open the way to its

experimental analysis" (Spemann and H. Mangold, 1924, p. 634). But a year earlier, in the Rektoratsrede, the address on the occasion of his election as the rector
of the university, he was more explicit:

> Driesch believes that he must renounce entirely a causal understanding [of the
> harmonious-equipotential system]. For him it is an ultimate given, not amena-
> ble to further analysis and thus one of his most important proofs of the auton-
> omy of life. I would leave it undecided whether, and how soon, we will encoun-
> ter ultimately unsolvable problems in the direction taken by Driesch. However,
> I believe that I can show that several further steps can be taken in the analysis
> of the harmonious-equipotential system. The method for this is the test of
> potentialities by transplantation. (1923, p. 13)

He then describes the transplantation experiments on the early gastrula and the
organizer experiment and continues, "With these findings, a breach seems to have
been made in the framework [Gefüge] of the harmonious-equipotential system
which was unassailable up to now—a breach through which experimental analysis
can penetrate further" (1923, p. 15).

I think that these comments of Spemann speak for themselves. They do not
bear out the claim that his interpretation of regulation and of harmonious systems
has a vitalistic aspect. On the contrary, he deserves credit for being one of the first
to point out the weakness of Driesch's argument and to have shown that the phenomena of regulation and of the harmonious-equipotential system are accessible
to experimental analysis. Of course, this is not more than a first opening. And we
have to admit that workers in the field have not made great progress in the last
six decades.

Nowadays, the simpler term "morphogenetic field" has replaced the old term
of Driesch. We find one of the first usages of this term in the repeatedly mentioned
postscript of Spemann's publication of 1921: "An organizer creates in the indifferent material in which it is located, or to which it is transplanted, an 'organization field' of definite direction and extent" (1921, p. 568). This was a year before
Alexander Gurwitsch (1922) and several years before Paul Weiss (1925) introduced the term "morphogenetic field" into the terminology of experimental
embryology.

Double assurance. Labile determination

Although Spemann was averse to the formulation of a general theory of development, theoretical issues were always on his mind. In the late 1920s, when I was
in Freiburg, the liveliest debates centered around problems of embryonic induction. In his publications of 1927 and 1931 he indulged in lengthy discussions of
this topic, which went far beyond the subject matter of the experiments.

When one refers to neural plate induction by the subjacent mesoderm, or to
lens induction by the optic vesicle, the misleading impression is created that one
is dealing with a single finite event. Spemann soon became aware that the actual
process of induction is very complex indeed. However, the road to the full clari-
fication of the problem was anything but smooth, and he did not live to witness
the experimental validation of all of his ideas. In his theoretical ideas he was
ahead of his time.

When he reexamined in 1927 the status of the problem of lens determination, he realized that he was confronted with a seemingly paradoxical situation:

> It is certain that the optic vesicle can induce lens in foreign ectoderm. On the other hand, it is likewise a fact that the lens-forming cells can differentiate also without an optic vesicle. At least in one form, *Rana esculenta,* both faculties undoubtedly occur side by side—one of which seems to be superfluous. Hermann Braus, using a term taken from Rhumbler, has spoken of double assurance in connection with another particularly clear case, discovered by him. (1927, p. 950)

(Rhumbler in 1897 had used the term in the context of cell division. Braus, a German anatomist, was a close personal friend of Spemann). What did Spemann have in mind, when he adopted the strange term? Double assurance (doppelte Sicherung) had been taken over from the terminology of engineers; they build into their constructions, such as bridges, safety devices that can support loads far in excess of normal demands. But how did a term with decidedly teleological flavor find its way into the realm of rigorous causal analysis? In fact, it never entered the vocabulary of general experimental embryology—and it soon became obsolete. Nevertheless, the term deserves its historical place; if nothing else, it was the springboard for the emergence of several very significant concepts. By the way, Spemann could have avoided the antiteleological misgivings, if he had substituted the term "synergistic principle" which he used synonymously with "double assurance" (see 1931a, p. 508; 1938, p. 92).

Originally, Spemann took the metaphor quite literally: one agent would duplicate the task of the other. In this sense, one of them would, indeed, be superfluous. But is the performance of the two agents necessarily identical? The original version was modified and refined when Spemann realized that the two agents might be complementary. "One factor can accomplish the same as the other; but a perfect end result is attained only by the combined action of both" (1931a, p. 508).

A decisive new step was taken when another dimension, the time element, was introduced: the two agents might operate consecutively, rather than simultaneously. One observes that the static nature of the metaphor "double assurance" gradually gives way to the demands of the uniquely dynamic nature of embryonic development. The idea of a two-step induction was conceived first in the context of neural plate determination: "It would be entirely conceivable that [its] induction which occurs after its exposure to the subjacent mesoderm is merely the continuation of another induction which was initiated when the [prospective mesodermal and ectodermal] materials were still lying side by side on the surface" (1927, p. 950). The last sentence has a familiar ring. We perceive now why the idea of neural induction by a two-step sequence had a particular attraction for Spemann. It permitted him to combine one of his favorite notions, his strong— and, as I have ventured to guess, emotionally or subconsciously motivated—commitment to the idea of a forward spreading of a neuralizing agent in the surface layer with the unquestionable evidence for neural induction by the subjacent mesoderm. He leaves no doubt about this point: "I have been familiar with the notion that the neural plate might be determined by a process that starts from the organization center at the upper blastoporal lip. I never gave up this possibility, when the other one, that is, determination by the subjacent mesoderm, had found an experimental basis in the results of A. Marx" (Spemann and Geinitz, 1927, p. 173). It does not matter that the notion of neural determination in the surface layer was never substantiated; what matters is the conception of the seminal idea that a sequence of several inductive acts may be required to stabilize the capacity of a group of embryonic cells for self-differentiation.

It was only natural to extend this notion to lens determination: "In the same way as the neural plate, the lens anlage is prepared, probably in connection with the neural plate, and specifically with the eye anlage contained in it, and perhaps induced by the latter. As a consequence of the ensuing folding processes, the optic vesicle is brought in contact with the lens anlage which is still more or less labile, thus completing its determination" (Spemann and Geinitz, 1927, p. 173–174). Two points here require further comment. In the early neural plate, the eye anlage is located in its anterior part, near the midline, and the future lens-forming cells arc located at the same level in the epidermis, at some distance lateral to the neural plate. These positions had been determined by vital staining. How could the prospective optic anlage induce the lens-forming cells across a considerable distance? Spemann was not quite clear as to the mechanism involved: "I believe that in *Rana esculenta,* for instance, the first lens anlage in the epidermis of the neurula originates somehow in connection with the eye anlage in the neural plate. But both [would seem to originate] as parts of a harmoniously segregating pattern and not by transmission of a substance on the part of the eye anlage which would determine specific epidermal cells at some distance from the neural plate to become lens-forming cells" (1938, p. 95). One can argue that if the transmission of a substance is considered improbable, one can hardly speak of induction. What he had in mind probably comes close to what we would now call self-organization of a morphogenetic field that would include both the future eye and lens anlage. But irrespective of the mechanism involved, the essential point is that lens determination, like neural plate determination, is envisaged as a two-step process. Here Spemann is again close to the mark.

The second point that deserves a further comment is the introduction of the concept of labile determination. The state of determination of the prospective neural plate at the beginning of gastrulation, before it is exposed to the inductive action of the mesoderm, is characterized as "still very labile" (Spemann and Geinitz, 1927, p. 173). The notion of "labile determination" introduces a new facet in the overall picture of embryonic induction. It suggests that in an early phase of progressive differentiation it is still possible to redirect the cells to proceed along a different pathway. Labile determination and the notion of induction as a two-step process are conceptually closely related. The first step in induction would result in labile determination. Apparently, the idea occurred to Spemann not in connection with a particular experimental result, but by theoretical reasoning, when he took a critical look at the merits and limitations of the method of transplantation. No doubt it had been eminently successful in his hands, but it had one weak point: in those instances in which the transplant underwent self-differentiation in its new location, the situation was clear; it proceeded with its program, uninfluenced by its new environment. However, when it differentiated in conformity with its new environment, the situation was equivocal. Either it had been completely uncommitted at the moment of transplantation, or it had been predisposed already toward its normal fate, but the course of its differentiation was still reversible. Spemann was quite emphatic on this point. "In all my publications and those done under my direction, there was always reference to 'relatively indifferent' or 'capable of redetermination'" (Spemann and Geinitz, 1927, p. 157).

I think that at that time Spemann did not fully realize the interesting implications of this assertion. The notion of redetermination was merely a logical deduction; it had no experimental basis and therefore no immediate significance. But the thought that differentiation, once initiated, could be reversed and redirected along a new pathway, was indeed intriguing. And in retrospect, the idea appears as the premonition of a significant and fascinating phenomenon. I refer to the discovery of *"transdetermination"* of imaginal disks of insects by Hadorn (1965). Imaginal disks are groups of undifferentiated cells in larvae that represent the primordia of adult organs. For instance, there are wing, leg, eye, and genital disks. Hadorn implanted such disks into the abdomen of adults—in his case *Drosophila* flies—where they grew and proliferated. After several weeks they were transferred to another adult fly. Hadorn made the startling discovery that when the transfers were repeated many times and disks were then implanted in an old larva and permitted to undergo metamorphosis, a certain percentage had changed their fate. For instance, genital disks would produce legs, wings, or antennae. This phenomenon was referred to as transdetermination. Another case in point which was discovered more recently is the *transdifferentiation* of embryonic retina cells to lens and pigment cells or of pigment epithelium cells to neural retina (Okada, 1980). In this case *differentiated* cells became redifferentiated. Thus the original idea of "double assurance," which seemed to be a teleological fantasy, can claim to be the progenitor of a very productive train of thoughts which contributed substantially to the elucidation of embryonic induction and related phenomena.

Do the concepts of a sequential order of inductions and of labile determination have any experimental foundation? I shall discuss first the scanty data that were available to Spemann and then take up some more recent developments. The issue of predetermination of the neural plate in the early gastrula was highly controversial in the 1920s. If predetermination exists in this early stage, it would be, at best, very labile, since Spemann and O. Mangold had shown that cells of this region can be readily induced to differentiate as epidermis or chordamesoderm. Spemann realized that a crucial test of self-differentiation would be to isolate parts of the prospective neural plate in tissue culture, but *in vitro* experiments by him and Geinitz were unsuccessful (1927, p. 950). In his laboratory, Lehmann, the guest from Switzerland, did extirpations of part of the organizer. The rationale of the experiment was that evidence for the predetermination of the neural plate would be forthcoming if the defects in the mesoderm would not disturb the normal formation of the neural plate. In most instances, however, mesodermal defects were accompanied by neural defects, and only a few cases could be interpreted as indications of predetermination (Lehmann, 1926, 1928). Spemann considered this possibility, but he was not entirely convinced (Spemann and Geinitz, 1927, p. 160).

In the meantime, much more forceful arguments in favor of neural predetermination were advanced by Goerttler, a student and colleague of the German anatomist Vogt. The design of his experiments was strongly influenced by Vogt's theoretical inclination toward a preformistic view of development, according to which organ primordia are predisposed toward their normal fates early in development. Goerttler had participated previously in Vogt's mapping experiments by vital staining; his contribution had been the mapping of the neural plate in the early gastrula, which turned out to be represented by a crescent-shaped area in the upper hemisphere just below the animal pole. He had described how large-scale wheeling movements during gastrulation bring these cells into their final position (Goerttler, 1925). Goerttler's first experiment was similar to that of Lehmann, except that in Goerttler's experiment the defects of the organizer material were so large that they interfered seriously with mesoderm invagination. Nevertheless, neural folds differentiated at the site of the crescent where they would be expected if the wheeling movements had been prevented. Goerttler concluded that since induction by the chordamesoderm seemed to be excluded, the prospective neural material must have possessed self-differentiation tendencies in the early gastrula (Goerttler, 1926).

In a second experiment, large defects in the prospective neural anlage of the early gastrula resulted in corresponding defects in the neural folds. In a third experiment, the prospective neural ectoderm of the early gastrula was transplanted to the flank of an older embryo, a neurula with rising folds. Under certain conditions, the transplants did form neural tissue. This again was interpreted as evidence for the predisposition of the transplants toward neural differentiation (Goerttler, 1927).

How did Spemann react to these provocative assertions of a much younger and rather brash colleague? He gave them serious consideration and devoted to them several pages in the discussion of the 1927 publication (Spemann and Geinitz, 1927, pp. 161–164). He pointed out several weak spots in Goerttler's experimental design and conclusions, such as the apparent presence of mesoderm underneath most of the supposedly self-differentiating neural folds. (Goerttler's descriptions were not always precise on this point.) Spemann acknowledged, however, that the third experiment, particularly, carried some weight. But he implied in a subtle way that Goerttler might have overinterpreted his data. At the end Goerttler gets a reprimand:

Goerttler says: 'Since in my case the presumptive neural anlage was capable of differentiation with great independence, one cannot quite understand what would be gained for an understanding of normal development, if one assumes a complicated mode of development when a simple action which experiments have demonstrated to be quite possible, can do justice to the phenomena in the same way' (Goerttler, 1926, p. 338). I gladly admit that this complicated mode of development is very curious . . . One can also understand the impatience with which one would like to cut through the knot . . . Yet there is no alternative to continuing patiently the work on the solution of the problem; and the first precondition for that is to let stand also the facts of the other side as they stand, and to acknowledge their proper significance. (Spemann and Geinitz, 1927, p. 164)

The work of Goerttler has not withstood the test of time. Years later, J. Holtfreter repeated two of his experiments and refuted all of Goerttler's claims. Holtfreter's experiments were far superior in execution and histological checks. In a now classical experiment, he reared early gastrulae in a hypertonic salt solution, which became known as "Holtfreter's solution," and obtained "exogastrulae." Instead of invaginating, endoderm and mesoderm evaginated and formed an inside-out axial system (endoderm covering notochord and somites) which had no ectodermal covering. It was connected by a narrow stalk with the hollow ectodermal vesicle, which showed no signs of neural differentiation (see Fig. 6–5). "In the absence of any mesodermal substrate [Unterlagerung] there was in all our exogastrulae not a single nerve cell, but also never a specific epidermal structure, such as typical skin epithelium, lens, balancer. Only ciliated and secretory cells were found" (J. Holtfreter, 1933f, p. 722). He concluded: "The neural inducing influence of the substrate [mesoderm] does not signify a double assurance; rather, it carries the full responsibility for the formation of the neural plate and all its derivatives" (*ibid.*, p. 723). Next, Holtfreter repeated on a large scale Goerttler's experiment of transplanting early gastrula ectoderm to the flank of older embryos. Holtfreter proved beyond doubt that the formation of neural folds by the transplants was the result of induction by the underlying somites and not self-differentiation. The evidence was absolutely convincing, since both prospective epidermis and prospective neural plate, transplanted to the same site, became neuralized (Holtfreter, 1933a). Finally, he succeeded where Spemann and Geinitz had failed. He reared prospective epidermis and neural material in Holtfreter's solution and managed to keep the explants alive for weeks. They formed epidermis but not neural tissue (J. Holtfreter, 1934a; see however, J. Holtfreter, 1945).

These brilliant experiments ended once and for all the claims of predetermination of the neural plate, and they also buried Spemann's favorite idea of the spreading of a neuralizing agent within the ectoderm.

The other system which preoccupied Spemann's theoretical reflections on labile determination, the lens and its determination, did not seem to fare much better. No further progress was made in the analysis of lens induction during his lifetime. He insisted that lens determination would conform to the model of a two-step induction process; but his specific suggestion that the initial step might occur in the neural plate stage was not subjected to an experimental test. Yet it was this system which, many years later, provided a triumphant vindication of the principle of the sequential order of steps in induction, which I consider to be one of his major contributions to analytical embryology. Of necessity, I shall give only a very brief and oversimplified account of what turned out to be a very complex story (see the reviews of Jacobson, 1966, and Twitty, 1955).

Harrison was the first to discover a two-step induction process. His experiment on induction of the otocyst (inner ear vesicle) in the American salamander, *Ambystoma punctuatum* showed that in the first step, during gastrulation, the mesoderm of the head region comes in temporary contact with the prospective ear ectoderm and conditions or predetermines this area for otocyst formation. Somewhat later, at the stage of rising neural folds, the folds of the hindbrain region establish contact with the same ectoderm area and complete the process of otocyst induction (Harrison, 1935, 1945). Harrison's student Chester Yntema made an additional significant contribution to this problem. In carefully timed transplantation experiments, he replaced the prospective ear ectoderm with prospective gill ectoderm, between the gastrula stage and the late neurula stage. He demonstrated that the two successive induction phases are precisely synchronized with two successive phases of reaction capacity (competence) to the respective stimuli. The gastrula ectoderm is responsive only to the mesoderm inductor, and the ectoderm of the neurula is responsive only to the neural fold inductor (Yntema, 1950, 1955).

The reinvestigation of lens determination started in the 1950s and led to the discovery that three inductive agents operate in succession. In the late gastrula, the endoderm of the future pharynx comes into temporary contact with the prospective lens ectoderm; during neurulation, the anterior edge of the mesoderm mantle, which is future heart tissue but then still located in a dorsal position, exerts its influence. Finally, in the tail bud stage, the retinal part of the optic vesicle completes the determination process. This implies, of course, that what appeared as self-differentiation of the lens epithelium overlying the optic vesicle in species such as *Rana esculenta* is, in fact, the result of preceding inductive actions, though the eye anlage in the neural plate, which Spemann had tentatively implicated as an inductor, was not among the inductors. These findings are the combined efforts of many investigators, prominent among them Antone Jacobson. His explantation and transplantation experiments gave conclusive proof of the three-step induction process (Fig. 3–6). To mention only one series: prospective lens epithelium was explanted in contact with either the endodermal inductor or the mesodermal inductor, or both. The evaluation of the data in terms of percentage of positive cases showed that the effects of the three inductors are additive; that their respective contributions differ in different species; and that "The endoderm and mesoderm are at least equal to the retina in importance as lens inductors" (Jacobson, 1966, p. 30).

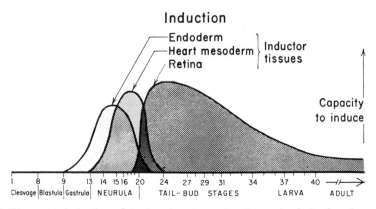

Fig. 3–6. Three-step induction of the lens in salamander. Successive inductions by endoderm, heart mesoderm, and optic vesicle. Note that the inductive capacity of each inductor waxes and wanes. From Jacobson, 1966.

The notion of multiple inductors operating *in tandem* can be elevated to a more general principle, since it was found that the nose and the otocyst (inner ear), which, like the lens, originate as epidermal placodes on the side of the head, are likewise programmed by a succession of three inductors. I have mentioned that, according to Harrison, head mesoderm and the neural folds of the hindbrain region function as otocyst inductors. Jacobson has observed an initial contribution of prospective heart mesoderm preceding the other two inductors. In the case of the nasal placode, anterior endoderm, prechordal mesoderm, and the rising neural folds at the forebrain region are involved as inductors. It should be stressed that in all three instances, the individual inductors operate only during a limited time, and the reaction capacity ("competence") of the placodal ectoderm is synchronized with the activity of the inductors, as was described for the otocyst induction.

In the light of these findings, Spemann's statement of 1927 that "one could imagine the entire course of amphibian development as a chain of doubly or multiply assured and causally connected single events" (Spemann and Geinitz, 1927, p. 174) is a remarkable premonition.

Hierarchy of organizers

The preceding quotation states in one sentence two different basic principles. The words "doubly or multiply assured" refer to the principle of sequential steps in the induction of individual structures which was the topic of the preceding section. The words "chain of . . . causally connected individual events" refer to an entirely different principle, that of a hierarchy of inductors. This principle was introduced by Spemann in 1924: "The induction of the neural plate by the underlying mesoderm reminds one of another developmental process, which has also been established in amphibian development, that is, the induction of the lens in the epidermis by contact with the optic vesicle. One can therefore designate the optic vesicle as the organizer of the lens . . . or an organizer of second order." (1924b, p. 1093). In other words, a structure which originated by induction becomes, in turn, the inductor for another structure.

This idea inspired an experiment that is characteristic of Spemann's experimental-analytical imagination. It is the kind of experiment on the living embryo that fascinated him more than the plan to test the inductive capacity of the devitalized organizer. The experiment is actually the combination of two experiments which had been done in his laboratory. O. Mangold had replaced the upper blastoporal lip with ventral ectoderm. The transplant invaginated and became mesoderm, thus demonstrating that in the early gastrula germ layer specificity had not been established as yet (O. Mangold, 1923). Marx, using the Einsteck-method (dropping the transplant through a slit into the cavity of the early gastrula), had shown that mesoderm of a late gastrula can induce neural plate in the ventral ectoderm of an early gastrula (Marx, 1925). In the breeding seasons of 1924 to 1926, Spemann and Geinitz substituted vital-stained prospective ectoderm for the upper blastoporal lip of an unstained gastrula. After the transplant had invaginated, it was removed and slipped into the blastocoele of a third gastrula. There it induced a secondary neural plate. Thus it was shown that the prospective ectodermal transplant had acquired inductive capacity along with mesoderm identity. In the discussion, Spemann writes:

> The organizing capacities [of the transplant] are hardly newly created but rather awakened and activated in its store of potentialities [Anlagenschatz]. Such a cell group can be called a 'secondary organizer,' or an 'organizer of second order'. . . . It is reasonable to generalize these results and to conceive of the total course of development, at least in amphibians, as a chain of successive, causally related partial processes, whereby always one embryonic part, while pursuing its own development, would, at the same time, give other parts the impulse to homologous or heterologous differentiation. (Spemann and Geinitz, 1927, p. 155)

It was shown later that the lens is involved in inducing the differentiation of the transparent cornea (Twitty, 1955), and the otocyst in the induction of the cartilaginous inner car capsule (Yntema, 1955). Spemann would have designated them as "organizers of the third order."

While there is no doubt about the validity of the basic principle of chains of inductions, the terminology, which makes "organizer" synonymous with "inductor," is untenable. It might have been acceptable in the early days, when the optic vesicle was considered as "the" inductor of the lens. But once the principle was established that inductions are complex, multiple-step processes, the ordering became meaningless. Spemann himself suggested another difficulty: that the "primary" organizer itself might originate by induction: "If one extends this notion [of chain of inductors] backward, then the anlage of the mesoderm in the early gastrula could be imagined to have been determined in connection with other parts perhaps starting from a narrow region from which determination spreads" (Spemann and Geinitz, 1927, p. 174). Indeed, Nakamura (1978) and Nieuwkoop (1969) have presented evidence that the chordamesoderm acquires its specificity very early, perhaps already in cleavage stages, by interaction of dorsal and ventral regions.

Fortunately, the ordering system of inductors has been forgotten, and the term "organizer" has survived in its original pristine meaning, as the originator of a secondary embryo. One can assert that the Spemann-Geinitz experiment has revealed no new basic insight; that it was "merely" a virtuoso experimental confirmation of a principle that had been documented by other experiments before. Furthermore, it could be pointed out that amphibian development is more than a chain of inductions. I think that Spemann was astute enough to agree to all of this. But he might have responded that it was worthwhile to have demonstrated in a single experiment the complexity of interactions between supracellular units, which are the essence of the epigenetic model of embryonic development.

Dialogue with Walther Vogt

The heading of this section is not intended to be taken literally. To be sure, there must have been lively discussions between Vogt and Spemann at scientific meetings, and I have probably witnessed some of them, but my memory fails me. What I mean to convey is that while all was harmony and consensus in Spemann's scientific family, at least until 1932, Vogt had a very independent mind, and his occasional dissent was perhaps not unwelcome.

Vogt is remembered as the originator of the *fate maps* of early amphibian embryos, which were constructed on the basis of extensive vital-staining experiments (Vogt, 1929b; Fig. 3–7). The maps delineate particular regions of the gastrula which are destined to become particular organs. They are projections of the pattern of embryonic structures onto the surface of the early gastrula. They are definitely not indicative of the state of determination of these regions. These maps became invaluable to experimental embryologists. While this work was undoubtedly his most substantial contribution, he has other achievements to his credit which assure him a prominent place in experimental embryology. The vital-staining experiments, which traced the fate of small areas of the blastula and early gastrula by continuous observation, gave him the opportunity to follow the morphogenetic movements of all parts of the embryo in great detail, through gastrulation and neurulation to the tail bud stage. He gave the first complete description of the complex gastrulation process, in which invagination is combined with convergence of lateral areas and other directional movements. One has to know the history of the controversy concerning this phenomenon, which extended over many decades, and some of the bizarre beliefs that were seriously discussed, to appreciate his achievement: he provided the definitive solution of this problem. Through this study, he became intrigued with morphogenetic movements in general. Of course, Spemann and others were aware of them; they had also recognized them as processes that were different from structural differentiation. But it was Vogt who elevated morphogenetic movements to an independent status, as a phenomenon worthy of investigation in its own right.

A third major interest of Vogt's concerned another basic issue that had been controversial since the early days of experimental embryology: the respective roles of early patterning, or "mosaic" development, and epigenetic features, such

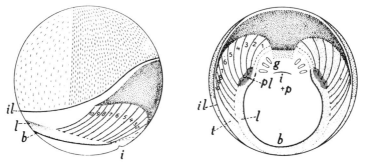

Fig. 3–7. Vogt's maps of the prospective organ areas projected onto the surface of the early gastrula of salamander, obtained by vital-staining experiments. Side (left) and posterior (right) views. *Dense broken lines,* Neural plate; *less dense broken lines,* epidermis; *coarse stippling,* notochord; *fine stippling,* mesoderm; b, ventral lip of blastopore; g, gill area; i, beginning of invagination (blastopore); il, limit of invaginating mesoderm and covering ectoderm; l, lateral mesoderm; p, vegetal pole; pl, pronephros (kidney) and limb area; t, tail region; 1–10, somites. From Hamburger, 1960, after Vogt, 1929b.

as regulation, inductions, and other interactions. Intuitively and by theoretical considerations, he was biased toward the idea of prepatterns, and he thought that the possibility of mosaic features had been unduly neglected by the Spemann school and epigeneticists in general. His imaginative ideas established him as a major figure in experimental embryology and as one of Spemann's peers, a colleague who could challenge him on basic theoretical issues.

I shall take up the last-mentioned problem first. The issue is well formulated in the title of his address at the meeting of the German Zoological Society in 1928: "Mosaic character and regulation in the early development of the amphibian egg" (Vogt, 1928). One notices that the two aspects of development are not considered as antithetical but as mutually compatible. He recognized fully the importance of the epigenetic mechanisms discovered and analyzed by Spemann and many others. Vogt wondered, however, whether the preoccupation with inductions, regulation, and organizers might not have brought about a neglect of less obvious but nevertheless important "mosaic" features in early development. In his address he gives a brief review of Spemann's work and states: "One would ask whether, besides this, there exist residual features [Restbestände] of a mosaic character in early development, that is, morphogenetic processes which are not mediated epigenetically but tied directly to a given embryonic region and thus prove . . . a primary capacity for self-differentiation [Selbstgestaltung]—where 'primary' would be related only to the fertilized egg as a starting point" (Vogt, 1928, p. 27). (Translations of Vogt are, of necessity, less literal than those of Spemann, since his prose is involved and interspersed with metaphorical expressions.)

This sounds like the replay of a record of the old controversy between Roux and Driesch in the 1880s and 1890s, when Roux insisted that each of the first two blastomeres of the frog's egg is endowed with a deeply engrained half-structure, whereas Driesch claimed with equal assurance that the two first blastomeres of the sea urchin egg are equipotential and that the regulative capacity of each blastomere to form a whole embryo reflects a quintessential property of early stages, that is, the absence of a prepattern of organization. Since then, a large number of experiments seem to have settled the issue: egg organization and regulation are not incompatible but are two aspects of early development which occur in parallel. Spemann has this to say in his book:

> The propensity for half-formation, as well as the capacity for formation of a whole, is present in both egg types [amphibians and sea urchins] in the two blastomeres. . . . The fertilized egg has a bilateral-symmetrical half-structure . . . which is transferred to the first blastomeres. If one blastomere is removed, an internal structural reorganization takes place in the other. . . . It results in the regulation of the half to form a whole, resulting in the formation of twin embryos. (1938, pp. 32–33).

Vogt was in essential agreement with this position, and he accepted induction and organizer action as an essential part of development. On what grounds, then, did he base his dissent from Spemann and the epigeneticists? He had serious reservations about the experimental method *per se*. The limitations inherent in the interpretation of isolation, defect, and transplantation experiments loomed large and, in his opinion, obscured rather than revealed the true nature of normal development. The undeniable fact that isolation experiments cannot tell us whether or not the first two blastomeres possess a half-structure, or that induction experiments cannot tell us whether a transplant was entirely uncommitted at the start or already predisposed toward its normal fate and then redirected, seemed to him to pose an absolute barrier to the understanding of normal development. This position, elevated to the level of a general theorem, implies that all experimental interventions involving, of necessity, a disturbance of the normal course of development, leave the central issue, that is, the nature of normal development, unresolved. Of course, every experimentalist is, or should be, aware of the antinomy inherent in his method; Spemann and experimental embryologists in general acknowledge it, but they assert that the insight gained by the analytical experiments outweighs by far their shortcomings.

Spemann himself went a step further:

> This [the epigenetic] mode of development is valid not only for the case of the experiment, that is, after disturbances; rather, it reveals a principle, according to which normal, undisturbed development likewise proceeds. The basis for this very important theorem is the principle of parsimony. If one can demonstrate a capacity of the embryo which suffices to explain the normal development, then one is not justified in assuming still another principle—unless facts are found which force one to do it. (1938, p. 34)

Clearly, Vogt and Spemann took opposing positions on this basic philosophical issue.

But Vogt did not want to let matters stand at an impasse. It is intriguing to see how he managed to have it both ways, that is, to design an experiment that would preserve the integrity of normal development and give new, and it was hoped, conclusive evidence for mosaic features in the egg. His solution was "to isolate partial processes of differentiation physiologically without changing the morphological continuity" (Vogt, 1927, p. 128). He did this in the following way: he constructed a chamber which was divided by a vertical partition made of silver foil. A salamander egg was fitted tightly in a hole at the bottom of the partition. A continuous flow of cold water was maintained in one compartment and a flow of hot water in the other. As expected, one half of the embryo developed faster than the other. Vogt referred to these embryos as "age chimaeras." He obtained curious combinations; in some instances, one half had developed neural folds, while the other was still in gastrulation. In other cases, a well-developed head with eyes and gills continued into an embryonic trunk and tail bud, or vice versa. Vogt interpreted the independence of development of the parts as clear evidence for mosaic features in the egg. He was careful, however, to avoid the notion of a pattern of preformed organ primordia; rather, he envisaged something akin to labile determination. He introduced the concept of facilitation [Bahnung]. He defined it somewhat differently from labile determination, which according to Spemann means an incipient, though reversible, predisposition toward normal development. Facilitation, as defined by Vogt, designates a specific reaction capacity, such as the capacity to react to a specific inductor (Vogt, 1928, p. 29). Years later, C. H. Waddington used the term "competence" for the capacity to respond in a specific way.

I think that this clever experiment did not provide the hard evidence for mosaic features that Vogt had hoped for. The only point he demonstrated was that each of the two first blastomeres can go its own way, even if it remains in physical contact with the other. But the evidence for half-structure is still indirect. Spemann pointed out in another context that direct evidence for egg structure cannot be obtained by the methods of experimental embryology, because transplantations and explantations of parts of the egg are not feasible (1938, p. 38; the transplantation experiments on egg cortex by Curtis, 1960, were discredited by Kirschner and Gerhart, 1981).

I believe that Spemann shared my criticism of the "age chimaera" experiment. It is never mentioned in his discussion of egg structure and mosaic development; perhaps he wanted to avoid rebutting a friend and colleague whom he respected. On the other hand, he was unwilling to ignore it altogether. He cited the experiment in an entirely different context, which was not intended or suggested in any of Vogt's writings. He quoted it as strong evidence against the application of Child's *gradient theory* to early amphibian development (1938, p. 336). The theory claims that gradients of physiological activity along the major axes of an egg (or of animals, like planarians, which are capable of regeneration) determine structural differentiation (Child, 1941). The theory presumes in eggs or embryos the existence of an anterior-posterior gradient with a peak at the anterior end and a dorsoventral gradient with a peak in the dorsal region. It predicts that the reversal of physiological activity along either one of the two axes will result in the reversal of structural organization. Assuming that physiological activity can be manipulated by temperature changes, the exposure of the ventral side to high temperature and of the dorsal side to low temperature in Vogt's experimental design should result in the shift of blastopore formation and subsequent neural plate formation to the ventral side. This, however, did not happen. In fact, Vogt did not observe any change in the normal disposition of embryonic materials in any of his age chimaeras (Vogt, 1928, p. 57).

I have mentioned already that Vogt strongly emphasized morphogenetic movements as processes separate from histological differentiation. This led to the question of whether these two major components of development were programmed independently of each other, and to the introduction of the terms "dynamic" and "material" determination (Vogt, 1923).

Both Vogt (1923) and O. Mangold (1925) had been impressed by the observation that transplants of the upper blastoporal lip expressed a strong tendency for invagination and stretching long before they showed signs of histological differentiation. But, of course, *post hoc* is not necessarily *propter hoc*. The only evidence that I could find for an independent determination of morphogenetic movements and structural properties is experiments in which parts of the prospective ectoderm of the early gastrula were transplanted to the prospective endodermal region (O. Mangold, 1925) or grown in isolation (Spemann, 1931a; J. Holtfreter, 1938a). In both instances, the pieces showed a strong tendency to expand and to stretch, without undergoing any kind of histological differentiation. This experiment does prove that dynamic determination (capacity to expand and stretch) can occur independently of material determination. A causal relationship between the two, however, has never been established. Spemann was probably right when he pointed out that it would be very difficult to obtain conclusive evidence, because the two processes are so intimately interwoven.

Vogt was twenty years younger than Spemann. He was handsome, aristocratic-looking, forceful and self-assured, perhaps a bit aloof. I insert here the reflections of my friend, Tuneo Yamada, now at the Swiss Institute for Experimental Cancer Research, near Lausanne. As a Ph.D. candidate he had spent several years with Vogt in Zurich and Munich, where Vogt was professor of anatomy. With his permission, I quote from a letter of January 1, 1985:

Vogt was not a person who would start right away to talk about his scientific opinions or discuss what was on his mind. He needed some introductory talk and a conducive atmosphere. But once he became interested, torrents of statements poured out which crystallized into ideas and arguments, while his eyes were fixed on the listener. After a while, he would stop and ask for the opinion of the listener, often with a characteristic smile. . . . He showed a talent for the formulation of concise sentences and emphasis on his points. His discussions were analytical and dynamic at the same time; they had style and fascinated the listener. . . . I do not feel that he was aloof. But he may have given this impression by his unwillingness to compromise and his preference for logical precision. This aspect, however, was amply balanced by his good humor expressed unwittingly, and his deep affection toward the people around him and toward nature. In my memory he appears as a smiling person, chatting amusingly and moving his arms vividly, rather than the stern professor, giving a lecture. . . . Living in Zurich and Munich, and being athletic, he went to the Alps very often for mountain climbing and skiing. Thus he became bound to the alpine nature. In my imagination, his attachment to the Alps, where Nature's formative force was so manifest, and to the formative movements in the embryo were coincidental. As a matter of fact, he seemed to be interested in the philosophy of morphogenesis, in general; I did not have a chance, however, to discuss this point with him.

Vogt and Spemann were coeditors of Roux' Archiv für Entwicklungsmechanik, *at that time the leading journal in this field. In this capacity, Vogt presented Spemann with a five-volume Festschrift of the* Archiv *on the occasion of Spemann's sixtieth birthday in June 1929. The address he gave when he made the presentation, which is printed in the first volume, is a wonderful testimony to Spemann's leadership and renown, but also to his personal magnetism. Vogt speaks of his "conquest of an uncharted territory of knowledge." He continues:*

The epoch-making success which gives your work commanding influence, not only in experimental embryology, but in the broader field of biology, is founded on your mastery of uncovering, in a new way, the coming into being of organizational patterns. . . . In particular, you have freed the vertebrate embryo on the one side from a sterile mosaic theory and, on the other side, from a vague epigenetic theory, and you have done this by the direct demonstration of specific potentialities of its parts.

And further;

Yet, something more personal is involved. The result is not sought by theorizing, thus passing over the object; nor by treating the embryo as a pure object, but rather by questioning it like a living partner, in a dialogue, as you yourself have expressed it once (Vogt, 1929a, xiv–xv).

I think that was the way most of us felt about Spemann's work.

The occasion of the sixtieth birthday was a two-day festivity which Spemann remembered later as one of the highlights of his life. A large number of his colleagues, present and former students, and friends were assembled. The auditorium in the zoology building was filled. The big table on which the demonstrations for his favorite lecture on birds were displayed, skeletons, feathers and all, was covered with flowers, and when he entered he quipped that he might as well lecture on botany. We then listened to a string of speeches, some solemn and others witty. Of course, it ended with a big dinner at which the local wines were appreciated more by the other guests than by the guest of honor. Spemann refrained from alcohol, and he never smoked. He had to guard his precarious health carefully.

On the evening before, a small, more intimate group was invited to his house. Among other surprises, F. Süffert showed a witty film which he had made. In it we younger colleagues and the Doktoranden (Ph.D. candidates) displayed our scientific exploits and other amusing episodes. I remember one scene that had been filmed on a pond near Freiburg: two students, fishing side by side, pulled a double-headed salamander out of the water; each had hooked one head. At the end of the film, one sees Ross Harrison, bent over the binocular microscope, doing an operation; then he rises, walks over to the blackboard, and writes in big letters "Yale congratulates."

The only discontents on this occasion were the subscribers of Roux' Archiv, and particularly the libraries. Without warning, they found themselves saddled with five extra volumes which cut deeply into their budgets. Their protestations compelled the publisher, Springer-Verlag, to vouch that henceforth a fixed subscription price would be strictly adhered to.

Vogt's untimely death of cancer, at the age of 53, was the occasion of an obituary by Spemann, in which he extolled his friend's achievements, his craftsmanship, his independent judgment, and his originality (1941). This was to be Spemann's last publication. He died six months later, in September of 1941.

Was the Term "Organizer" Appropriate?

In the minds of many, the organizer epitomizes the greatest achievement of developmental biology in the early part of this century; others have questioned the validity of the term almost from the beginning. I intend to subject the organizer concept to a critical evaluation, not from the present-day vantage point but from the perspective of the late 1920s and early 1930s. This I find necessary for historical reasons. The situation changed profoundly after the discovery in 1932–33 that dead organizers and tissues retain the capacity to induce complex structures. In a later chapter I shall ask how the organizer managed to survive this crisis.

I have seen many induced embryos, and quite a few were more complete and impressive than the best case in the material of H. Mangold (Fig. 3–4). My immediate reaction, which I shared with many others, was that the small piece of the upper blastoporal lip which creates a harmoniously organized embryo fully deserves the designation of "organizer." But on second thought I wondered: what is really meant by "organizer" and "organizing"?

The operational definition of the organizer states that it is a piece of the upper blastoporal lip of the amphibian gastrula which, when transplanted to a relatively indifferent region, causes the formation of a secondary embryo. Since this definition seems to be unassailable, why did the term give rise to so many misunderstandings and misgivings? A simple answer would be that the somewhat pretentious term suggests a single agent or principle that creates organization in an otherwise unorganized material. "It would be misleading to conceive of the organizer material as a kind of general manager which determines the destiny of the entire remainder of the embryo" (J. Holtfreter and Hamburger, 1955, pp. 279–280). Or, in the words of Harrison, the preeminent arbiter in these matters: "The use of the term 'organizer' is likely to be attended by some confusion, for the word may be readily taken to imply more than we are really justified in attributing to the thing itself" (Harrison, 1933, p. 317). On another occasion, he stated, "Since the word 'organizer' connotes a master regulator which creates organization, and since there are in the course of development many actions of the same general character, that could hardly be accorded such a role, it is perhaps more appropriate to use the term 'inductor' to denote processes of this kind" (Harrison, 1969, p. 29). Indeed, terms like "primary inductor" or "determiner" would have avoided the stigma of the word "organizer." They would have failed, however, to denote one of its essential functions, that is, to create an integrated whole embryo out of heteroplastic parts. However, it should be stated in fairness to Harrison that he realized this: "Organizers have come into prominence through the dramatic manner in which they have demonstrated epigenetic development at a time when the tendencies of thought were in the direction of preformation. Their most striking action, still veiled in mystery, lies not in the induction of a particular organ, here or there, but in making plastic material form a harmoniously constructed embryo" (Harrison, 1937, p. 373).

In the following I shall marshal first the factual experimental evidence, and then conceptual arguments that invalidate the notion of the organizer as a master regulator and place it into the proper perspective.

Limits of the jurisdiction of the organizer

It was mentioned in the discussion of Spemann's early constriction experiments that a frontal constriction separates a ventral "belly piece" from the dorsal region, including the organizer. The belly piece was found to contain mesoderm and endoderm and occasionally kidney tubules and red blood cells (1902). In accordance with these findings, Spemann, Vogt, and O. Mangold observed an autonomous invagination of the lateral and ventral regions around the lateral and ventral blastoporal lips, when they were separated from the dorsal regions (discussion in Spemann, 1938, p. 101). Hence both morphogenetic movements and certain differentiations can proceed in ventral regions independently of the organizer.

Spemann's publications contain few references to the endodermal derivatives.

Later *in vitro* experiments have shown that the anlagen of the different sections of the intestinal tract in the endoderm of the early gastrula are capable of region-specific differentiations (J. Holtfreter, 1938a). It is true that the fate of these parts is not yet irreversibly fixed in the gastrula stage but changeable under experimental conditions such as combination with different mesodermal derivatives (Takata, 1960); this, however, is not relevant to our main argument: that the patterning and differentiation of an important organ system is not subject to the jurisdiction of the organizer. The activity of the organizer is restricted essentially to the dorsal parts of the embryo.

Interactions of the organizer and adjacent ectoderm

The following experiments challenge the conventional view that the organizer is irreversibly programmed for invagination and chordamesoderm differentiation. A student of Vogt, E. Bruns (1931), removed a large portion of the roof of a salamander blastula, that is, prospective neural plate and epidermis. After the healing of the wound, a normal neural plate developed which was proportionate to the smaller overall size of the embryo. What is remarkable is not that regulation occurrred, but that the neural plate was formed by material bordering the cut edge that would normally have formed chordamesoderm. Obviously, part of the organizer was prevented from invagination by the stretching required for the closing of the wound. Another part of the chorcamesoderm must have invaginated; otherwise no neural plate would have been induced. In other words, material that was already programmed for mesoderm formed an ectodermal derivative, and furthermore, this "ectodermization" of mesoderm was linked to the prevention of its invagination. The experiment was repeated later on a larger scale by Holtfreter. He removed in the early gastrula the entire prospective neural plate region, as delineated in Vogt's fate map (Fig. 3–7), and obtained the same result. The derivation of the neural plate from organizer material was ascertained in both investigations by placing circular vital-stain marks at the cut edge of the chordamesoderm; they were found later as elongated parallel stripes on the neural plate. Holtfreter comments: "The result filled me with amazement. As soon as the enormous defect had healed over, a normal neural plate was formed, and a normal embryo developed, endowed with all sense organs" (J. Holtfreter, 1938a, p. 608). The experiments leave no doubt that chordamesoderm can be diverted from its normal fate and produce what is normally an ectodermal derivative. Paradoxically, as newly created ectoderm, it subsequently comes under the inductive influence of the very structure of which it was originally a part.

If the ectodermization of mesoderm needed further proof, it was provided by isolation experiments performed by Holtfreter. He accomplished the feat of growing the organizer, or parts of it, in a balanced salt solution for several weeks, during which the explants underwent complete tissue differentiation. He found that, in addition to the expected notochord and muscle, the pieces had produced unmistakable neural tissue in 70%, and epidermis in 40% of the explants. The neural tissue usually formed brain vesicles (J. Holtfreter, 1938a). This transformation is reminiscent of a similar phenomenon called "transdetermination," which was observed in transplants of imaginal disks of insects (Hadorn, 1965).

While the ectodermization of mesoderm remains unexplained—it is not even clear whether the mechanism involved is the same in the defect and explantation experiments—this phenomenon has important implications. It reminds us again that it is an oversimplification to attribute to the organizer complete, unrestricted authority, both with respect to its own differentiation and its domination over the ectoderm. In fact, the organizer and the adjacent ectoderm interact with each other in a subtle way. In normal development the organizer can operate only if its inherent capacity for ectodermal differentiation is repressed and if its invagination tendency is not obstructed. These are significant constraints. Spemann was aware of the ectodermization of mesoderm (1938, p. 137), but he did not place this phenomenon in the same context as I have done.

Organizer action in normal development

One does not deprecate the extraordinary performance of the organizer in the H. Mangold experiment if one takes a closer look at its more modest role in normal development. One of its striking activities in the experiment is its capacity for assimilative induction. On further reflection, one realizes, however, that in the course of normal development, there is no need for this kind of induction. The chordamesoderm region, which in Vogt's map of the early salamander gastrula is represented by a crescent-shaped transverse band above the blastopore, has the properties of a self-contained field, which during and after neurulation differentiates into notochord and two rows of somites. It contains all the material required for the formation of these structures and does not have to recruit cells from beyond its borders. Spemann was aware of the apparent paradox that the chordamesoderm possesses inductive capabilities that are not materialized in normal development: "The situation is remarkable, indeed. A piece from the middle of the prospective mesoderm is charged with the capacity to force adjacent tissue to differentiate along the line of its own differentiation. But normally it is not required to make use of this faculty because the adjacent [lateral] region proceeds already along the same line and even has the capacity to impose its differentiation tendency on the middle region. Why this superabundance?" (Spemann and Geinitz, 1927, p. 171). Spemann had no answer; but the question should be pursued further, because it goes to the core of the dispute about the significance of the organizer. If the performance of chordamesoderm in normal development does not go beyond invagination, self-differentiation, and induction of the neural plate, then most of its glamor is gone; it is just another inductor and hardly deserving of special attention and a special name—and a Nobel Prize! Moreover, this is not the only reservation that can be advanced concerning the organizer concept.

Does the organizer represent a new principle?

There is general agreement that major components of organizer activity can be interpreted in terms of principles that were known in the 1920s and 1930s. This holds for the capacity of upper blastoporal lip material to invaginate autonomously and to self-differentiate into notochord and two rows of somites, and furthermore for its capacity of assimilative induction of host mesoderm and neural induction by contact. Although none of these processes were then—or later—

really explained in terms of the mechanisms involved, they had parallels in other systems that had been explored previously. The main feature of the organizer experiment, however, the integrated nature, or wholeness, of the secondary embryo, and its composition of chimaeric parts, seemed to pose a problem not encountered before. At least this is the impression one gets from some statements of Spemann. In one of the first reports on the organizer he says, "Such a secondary embryonic anlage looks as if it were built by a superior [übergeordnete] force out of material that happened to be available, without consideration of its origin or species affiliation" (1924b, p. 1093). In his book he states even more pointedly: "The organizing force which fashions the whole, overriding the boundaries of the [heteroplastic] material, poses ultimate questions of the developmental process, the answers to which now seem to be barely in the realm of possibilities" (1938, p. 147). Yet, at closer inspection, one realizes that we have encountered this phenomenon before, in the harmonious-equipotential system or morphogenetic field. The general definition of the field in terms of an embryonic unit which is self-differentiating as a whole but regulative within its boundaries is applicable to the prospective chordamesoderm of the upper blastoporal lip. The new and complicating aspect is not that small transplants from this region can regulate, but that, in addition, they complement themselves by assimilative induction of host material and also induce the neural plate. The fact that we are dealing with an extremely complex situation does not imply that a new principle is involved. And as far as a "superior force" and "ultimate questions of development" are concerned, Spemann's own contention has to be remembered that there is no mystique about the harmonious-equipotential system, but that it is amenable to experimental analysis. In fact, he had claimed that he himself had made the first step in its analysis by identifying the organization center as the starting point of its further differentiation (see above, Chapter 3, Spemann's Theoretical Evaluation of the Organizer Experiment: Breach in the Harmonious-Equipotential System).

What Spemann apparently found difficult to comprehend was that the secondary embryo was composed of irregular patches of material derived from two different species. Yet he himself had shown in his first heteroplastic transplantations that pieces of prospective ectoderm, when transplanted early enough to the region of the prospective neural plate, are incorporated smoothly in the neural tube; in one of his pictures, the walls of the forebrain and one optic vesicle are composed in part of pigmented *(T. taeniatus)* and in part of unpigmented *(T. cristatus)* cells (Fig. 2–16). In both instances, a morphogenetic field—in the latter instance, the forebrain-eye field—has undergone self-organization with disregard of the cellular composition of the material. This observation merely reflects another general tenet which holds for all vertebrate development: that self-organization follows the rule of a hierarchical sequence of structuring, in the sense that larger supracellular units become subdivided into smaller more and more specialized subunits until every part of the embryo has attained its final instruction. This implies that in vertebrate development, in contrast to that of annelids and molluscs, redundancy of cells is the rule, and individual cells are relegated to the role of anonymous bricks. This does not prevent them, however, from expressing their species-specific characteristics, such as size or pigmentation. The principle of self-organization is not flawed by the fact that we have a poor understanding of the mechanism involved. Indeed, modern ideas of patterning and positional information are still trying to elucidate this phenomenon, which has puzzled embryologists ever since Driesch discovered it almost a century ago.

What makes the organizer experiment unique is not the manifestation of a new principle but the unique constellation of important events at a critical period: the integration of self-differentiations, inductions, regulations, and self-organization, not just in the generation of "this or that organ" but in the creation of the *axial organ system,* and the unfolding of these activities during a critical relatively short time span, the gastrulation process.

The organizer in the context of early development

While the organizer experiment dramatizes a particular phase of early amphibian development, one has to be careful not to isolate this episode from the continuum of early development. Embryologists have always held to the idea that wholeness is manifest already in the egg and that the epigenetic events that transform the egg into the embryo operate within this framework. Child (1941), the founder of the gradient theory, was particularly emphatic on this point. A modicum of organization is present already in the unfertilized egg, in the form of animal-vegetal polarity which is discernible in most amphibian eggs in the darker pigmentation of the animal pole region. Fertilization initiates cytoplasmic movements resulting in the formation of the so-called gray crescent (not visible in all amphibian eggs), which is located opposite the sperm entrance point. The sickle-shaped gray crescent demarcates the future dorsal side of the embryo; hence, two further landmarks of organization, bilateral symmetry and dorsoventral polarity, are now established. In the subsequent cleavage of the egg and the formation of the hollow blastula, organization proceeds further, in that the chordamesoderm acquires its specificity. The organizer experiment has highlighted the crucial importance of the subsequent phase, gastrulation, which includes the invagination and initial differentiation of chordamesoderm and leads to the induction of the neural plate. However, these processes are embedded in a current of events, dynamic in nature, which begin in the blastula and continue without break through gastrulation and neurulation to tail bud formation (Vogt, 1929b). From the blastula stage on, every part of the embryo, and not just the upper blastoporal region, is involved in a sequence of continuously changing morphogenetic movements. Only a small area of the animal pole remains static. Vogt's vital-staining experiments revealed the complexity of these movements, which include expansion, longitudinal stretching, wheeling, foldings, invagination, and, in later stages, evaginations. They are as precisely integrated and synchronized as inductions. It is true that gastrulation plays a special role in the overall process. First, the three germ layers are created during gastrulation. The entire ventral hemisphere is invaginated into the interior; during this process, ectoderm, mesoderm, and endoderm are separated from each other, and they assume their relative positions with respect to each other. Second, the gastrulation movements bring the chordamesoderm in juxtaposition to the overlying ectoderm, a prerequisite for neural induction. This is a special case, illustrating a general rule: morphogenetic movements are an integral part of inductions, in that they establish the contact between the inductors and the tissues that are to be induced. Other examples are the above-mentioned three-phase inductions of the lens, otic, and nasal placodes.

My intent here is to place the organizer in proper perspective. The integrative activity of the organizer is only one in a series of organizing principles that operate in early vertebrate development, though a particularly interesting one. Furthermore, following the line of thought of Vogt, I stress once more the crucial relationship between dynamic processes and structural differentiations. Of the early events in vertebrate development, self-organization of morphogenetic fields seems to be the only process that does not require the mobility of cell assemblies.

A critical assessment of the organizer concept

If the organizer is not a master regulator, if its role in normal development is limited, and if no new principle is involved in organizer action, how can one understand the dominant position which the organizer concept held in the minds of the experimental embryologists of the 1920s and 1930s—and its survival to this day? To answer this question, the issue must be examined from a broad conceptual and historical perspective. Obviously, the term "organizer" is related to the terms "organism" and "organization." In the hierarchical order of life, from populations to molecules, the individual organism assumes a unique position. By definition, organisms are entities characterized by an organizational plan in which axial polarity, symmetry, and the patterned arrangement of organs are the points of reference. This is the axiomatic basis of experimental embryology and of all other branches of biology dealing with individual organisms. It is the special concern of experimental embryologists to identify the causal factors and developmental interactions that bring organization into existence, by application of the experimental method. Although Spemann, along with Driesch and others, preferred the term "developmental physiology" to Roux' term "developmental mechanics" (see the title of Spemann's first experimental studies), the focus was not on physiological events in the narrower sense of the term, but in the origin of form, that is, morphogenesis. This view of organization dates back to Aristotle and other earlier thinkers; certainly it is alive in the poet and naturalist, Johann Wolfgang von Goethe (who coined the term "morphology"), from whom Spemann derived much inspiration. Goethe spoke of "gepraegte Form, die lebend sich entwickelt" [the minted form that lives and in living is developing]. Since experimental embryologists were primarily concerned with the origin of the basic plan and the analysis of the origin of organs, that is, supracellular units, they had remarkably little interest in the role of cells (except for blastomeres) and only a limited interest in the role of the nucleus (see Hamburger, 1980a). Much later, Holtfreter became one of the pioneers of the study of cellular events in this field.

What Spemann had in mind when he coined the term "organizer" was a precise designation of that region of the vertebrate gastrula which had revealed in the transplantation experiment of H. Mangold the capacity to initiate the formation of an axial system. This discovery gave the answer to the quest which—as I have stressed repeatedly—motivated his experimental pursuits from the beginning: to come to an understanding of the causation of the axial organs which to him and to others epitomized the basic organizational plan of vertebrates. From this viewpoint, the choice of the term "organizer" cannot be faulted. But it is true that the organizer, in the experimental situation, delivered in some respects more, and in some respects less, than the operational definition implied. More, in that later, more successful, repetitions of the H. Mangold experiment produced more complete secondary embryos with heads, including mouth accessories, gills, balancers or suckers, trunks with viscera, and tails with fins. Less, in that in normal development the activities of the organizer are integrated in a continuum of equally important dynamic activities that transform a single cell into an organism. But nothing can detract from the fact that the secondary embryos represent an integrated whole, their integrated character being accentuated by the chimaeric structure of their parts. Stressing the holistic aspect of the secondary embryos is merely another way of restating the underlying axiomatic basis of the concepts of organism and organization. In my opinion, the term "organizer" is justified, if all the caveats that I have listed are heeded. No other term comes to mind that would characterize the results of the H. Mangold experiment equally well.

The holistic view was by no means exclusive to Spemann. When Harrison, at the same time (1921), devoted his major experimental efforts to the analysis of the origin of bilateral symmetry and laterality in the amphibian embryo, he implicitly accepted the same axiom.

It would be appropriate to end this discussion by placing the organizer and its discoverer in the broader context of contemporary experimental embryology. This task, however, would transcend the scope of my project. No doubt, the discovery of the organizer and the invention of the imaginative term were the most visible events in that period; they contributed greatly to the prestige of experimental embryology. This field now assumed a commanding position in biology and it extended its impact to other fields. For instance, gestalt psychology, which flourished at the same time, found concepts like regulation and patterning relevant to its pursuits. But it would be unfair if one did not give credit to other major accomplishments in experimental embryology. Great strides were made in invertebrate embryology, such as the cell lineage studies of E. G. Conklin and F. R. Lillie, and the masterful analysis of sea urchin development, begun by John Runnström and culminating in the brilliant experiments which Sven Hörstadius performed on these tiny eggs (references in R. Watterson, 1955). Major contributions of E. B. Wilson and C. M. Child also deserve mentioning. Spemann's leadership position was never in doubt. He shared it with R. G. Harrison, who was his peer in experimental accomplishments, perspicacity, and wisdom. Those of us who are familiar with his work and were inspired by him wished that he had shared the Nobel Prize with Spemann. This was also the opinion of Spemann himself. In a letter to me, dated November 20, 1935, in which he acknowledges my congratulations for the Nobel Prize, he writes: "The photo which you enclose was taken two years ago in New York. At that time, I also had my photo taken together with Harrison. We called it 'Max and Moritz' [after two comic characters in a famous German children's book of the same name by Wilhelm Busch].[1] I thought that perhaps we would later, together, 'arm-in-arm challenge our century' [quoted from Friedrich Schiller's drama 'Don Carlos']. I would have found it more appropriate [Ich hätte es richtiger gefunden]."

Actually, Harrison had come close to receiving the Nobel Prize. "In 1917, a majority of the Nobel Committee recommended that the Prize should be given to him 'for his discovery of the development of the nerve fibres by independent growth from cells outside the organism'. The Faculty, however, decided not to award the Prize for that year. When Harrison's work was again submitted to a special investigation in 1933, opinions diverged, and in view of the rather limited value of the method and the age of the discovery, an award could not be recommended (Nobel Foundation, 1962, p. 259). As I have stated elsewhere, "What was actually of limited value was the judgment of the Committee and not Harrison's achievements" (Hamburger, 1980b, p. 611).

---

[1] The photo is reproduced in Horder and Weindling (1986).

CHAPTER 4

Further Exploration of the Organizer and of Induction (1925–1931)

Organizers in Fused Eggs

I have discussed the contributions which the discovery of the organization center had made to an understanding of double monsters in vertebrates. O. Mangold added a new twist to this story. Spemann's theory of the origin of duplications was based on experiments on the fusion of half-gastrulae. O. Mangold accomplished the difficult feat of fusing two whole eggs in the 2-cell stage. He took advantage of the fact that when the vitelline membrane around the egg is removed, the egg flattens and the two cells move apart temporarily, assuming a dumbbell-shaped configuration. If at that moment another similar dumbbell-shaped egg is placed across the handle of the first, the four cells will fuse and form an oversized embryonic mass. Since in salamanders the first cleavage plane is either median, dividing the region of the future organization center in half, or frontal, leaving the organization center in only one (the dorsal) blastomere, it was predictable that different combinations would produce one, two, three, or even four axial systems. The experiments confirmed the prediction. When two eggs were fused in which the cleavage planes had separated dorsal from ventral, then two of the four cells contained the precursor of a whole organization center and the result was the formation of two axes. When two eggs were fused, one of which had cleaved in the median plane and the other in the dorsoventral plane, then three axes were formed, one from the dorsal cell which contained the precursor of a whole center, and the two others from the two cells which contained one half-center each, the latter being capable of regulation to form two complete axes. When both eggs had cleaved in the median plane, then each of the four cells contained the precursor of a half-center, and four axes were formed in the rare cases in which each half-center had managed to regulate to form a whole. Finally, a

90

single, double-sized embryo was formed when the precursors of one complete and two half-centers were adjacent to each other and fused to form a single center (O. Mangold, 1920). Later, Mangold and Seidel (1927) succeeded in fusing 2-cell stages from two different species (*Triturus taeniatus* and *T. alpestris*). The eggs of *taeniatus* have a lighter brown pigment than those of *alpestris;* hence it was possible to give a visible demonstration of the composite nature of some of the axial systems. Although these experiments cannot claim to have yielded new conceptual insights, they demonstrated conclusively that a precursor of the organization center is prelocalized in the dorsal region of the egg.

Homeogenetic Induction

While the holistic aspect of the organizer attracted wide attention, a specific feature, neural induction, assumed a central position and replaced lens induction as the paradigm of induction. Eventually, neural plate induction became the testing ground for the analysis of the mechanism of induction. In the meantime, Spemann (in Freiburg) and Mangold (in Dahlem) made a startling discovery which seemed to confound the current notion of induction. The results were reported in a joint publication (O. Mangold and Spemann, 1927).

They both had the idea of transplanting a piece of the anterior neural plate into the blastocoele of an early gastrula (Einsteck-method), though for different reasons. They expected the transplant to form a brain and optic vesicles, and they wondered whether the latter would induce lenses in the host ectoderm. Mangold was interested in knowing whether the age difference between the donor (neural plate stage) and the host (early gastrula) would interfere with lens induction. Spemann's intent was to create another "secondary organizer," using the same experimental design as in the previously reported experiment with Geinitz. He implanted a vital-stained piece of prospective epidermis to the region of the anterior neural plate where it was induced to form neural tissue. This piece was then cut out and implanted into the blastocoele of a third gastrula. If it formed an optic vesicle, and the optic vesicle induced a lens, then the original prospective epidermis would have been transformed into an organizer of second order. For unknown reasons, the implants formed mostly irregular brain parts and only a few incomplete eyes with no lenses. But the failure of the original plan was compensated by the unexpected finding that the implanted neural tissue had induced rather conspicuous neural tubes and brain parts. In the breeding season of 1928, Mangold extended the scope of the experiment: he tested the brains of older embryos. Some transplants did produce eyes which then induced lenses, thus answering his question. But more importantly, he found that brain tissue of tail bud stages and even of swimming larvae had retained the capacity for neural induction (O. Mangold, 1929). The terms "homeogenetic" or "assimilative" induction were introduced to designate this kind of induction, where "like begets like." (The term "assimilative induction" had been used before to denote a particular aspect of the organizer experiment: the assimilation or incorporation of host mesoderm into transplant mesoderm. In the present case, inducer-like structures were induced by contact. To avoid confusion, the term "assimilative" should be restricted to the former and "homeogenetic" to the latter type of induction.)

The new findings were unsettling. Never before had anybody encountered an induction which had no place in normal development. The normal brain is a heterogenetic inductor of nose, lens, and ear placodes. The homeogenetic induction of more brain tissue would not have made the larva smarter but anomalous instead. Spemann was ostensibly puzzled. "These facts cannot be readily incorporated in hitherto known facts. For that very reason, they promise to lead to an important extension of our knowledge" (1927, p. 949). His optimism was unfounded, however. Neither he nor Mangold came up with a new idea of how to proceed from there, and the experiment fell into oblivion. I believe that their failure to exploit their discovery was due to a too narrow, tradition-bound perception of what they had found. They placed the major emphasis on the structural identity of the inductor and the induced tissue. If, instead, they had stressed the atypical and abnormal aspect of the phenomenon, the idea might have occurred to them that perhaps the brain is not the only abnormal inductor of neural tissue. They then would have anticipated by several years Holtfreter's even more startling discovery that a large number of animal tissues can induce neural tubes (J. Holtfreter, 1934c).

However, Spemann did make an interesting comment which showed that to him the notion of a chemical inducing agent was more than an abstract principle. He depicted it in very concrete terms: "To explain this fact [of neural induction by neural plate], the simplest assumption would be that the neural-inducing stimulus which had originated in the archenteron roof [chordamesoderm] and had been taken up by the ectoderm can be released again by neural plate. This would presuppose that it [the agent] is not used up and does not disappear in [the process of] the determination and differentiation of the neural plate" (Mangold and Spemann, 1927, p. 358). Commenting on Mangold's new discovery, he states: "The long retention [of the agent] is truly remarkable; it is still present in the functional brain of a swimming larva. One can say that a residue of the neural-inducing agent is preserved and it can have an inducing effect on indifferent epidermis. This seems to speak in favor of the chemical nature of the agent" (1927, p. 949). Although this idea remained in the realm of speculation, it is clear that in the late 1920s Spemann came close to a realistic consideration of the chemical nature of induction.

Finally, an amusing and rather weird idea, dating back to the early days of lens induction experiments, deserves to be rescued from oblivion. Lewis, one of the pioneers in this field, suggested that the optic vesicle, while being transformed into an optic cup, might pull the overlying ectoderm inward mechanically: the lens vesicle would originate by some kind of suction (Lewis, 1907). In his definitive publication on lens induction, Spemann took this notion quite seriously. He collected all the arguments against it and ended with the sensible suggestion that "the stimulus is more specific and perhaps chemical in nature" (1912a, p. 90).

CHAPTER 5

The Mechanism of Induction:
The Turning Point (1931–1932)

The First Inductions by Disarranged Organizers

Although the question of the nature of the inducing stimulus had been on Spemann's mind for many years, it had not been subjected to an experimental test. Now its time had come. Characteristically, he approached the problem from his particular vantage point, assigning to structure a crucial role. By structure he meant polarity (longitudinal structure), bilateral symmetry, and regional structure, that is, regional inductive specificity. Consequently, he inquired whether structural integrity is a necessary prerequisite for inductive capacity. This question was raised for the first time in the organizer paper: "Concerning the means [Mittel] of the determinative influence, no factual clues are available as yet. Experiments . . . of crushing the organizer, whereby its structure would be lost, and then implanting it between the germ layers [into the blastocoele], might lead us further into this subject" (Spemann and H. Mangold, 1924, p. 634). He is more specific in a later review:

> We are completely uncertain about the most elementary question: whether induction is mediated *materially* or *dynamically,* that is, whether particular [chemical] substances or particular [physical] forces awaken the latent capabilities in the parts that are to be induced. If the latter is the case, then one would expect only living cells to be capable of induction. Particular substances, however, could be retained and remain effective even if the living substance which had produced it was destroyed, perhaps by freezing or crushing. Such experiments are being prepared. (1927, pp. 948–949)

Thus Spemann can claim priority for raising the question of the mechanism of induction and, more specifically, for suggesting an experiment that would indicate whether a chemical agent is involved. But he was in no hurry to fulfill his promise to do the experiment; it was postponed until 1929 and not reported until 1931. Why this delay? The situation must be viewed in historical perspective. Today, the search for an inductive agent would be high on our agenda and we would implicate a chemical agent as a matter of course. One can list several entirely credible reasons for Spemann's reluctance to move ahead. First, the time was not ripe for an aggressive chemical approach. Appropriate methods were just beginning to be developed, and the experimental embryologists would have been hardly aware of them, since there was little, if any, communication between them and their colleagues in what was then called physiological chemistry. This, however, did change in the 1930s. Second, Spemann faced a conceptual difficulty: his strong emphasis on structure was a hindrance to thinking in physiological terms. In the two paradigmatic cases, neural induction by mesoderm and lens induction by the optic vesicle, a self-differentiating structure initiates the differentiation of another structure which is qualitatively different from its own. It was not immediately obvious that a chemical agent could perform this feat, unless the structure was preprogrammed in the reacting tissue and the inductor had merely a triggering function. On this score, however, the data were ambiguous. On the one hand, there was evidence that inductions can operate across borders of species and genera, and even between salamander and frog embryos; this would indicate that, indeed, the stimulus is nonspecific. On the other hand, the discovery of the head- and the trunk-organizer phenomenon had shown that induction is instructive, at least in the limited sense of regional specificity, and the induction of a lens in flank ectoderm also implies that the inductor, the optic vesicle, has instructive capacities. This unresolved problem led to an impasse in the search for a chemical agent.

But I think, more than anything else, Spemann's personal inclinations motivated his course of action: his undisguised preference for experimentation on the living embryo always prevailed (see Hamburger, 1969). The organizer experiment had opened a cornucopia of new opportunities for indulging in this passion. Moreover, circumstances had prevented him from doing his own experiments over a long period, from 1919 to 1924. Finally, in the spring of 1925, he returned to the workbench with renewed vigor. He started two projects: the experiment with Geinitz on "secondary organizers" and an experiment with Mrs. E. Bautzmann (the wife of H. Bautzmann) on regulative properties of the organizer (Spemann and E. Bautzmann, 1927). No sooner were they completed, when two others were begun in 1927: the one which resulted in the discovery of homeogenetic induction and the very time-consuming series of experiments that led to the discovery of head- and trunk-organizers. The latter involved several hundred experiments. No wonder that the destruction of the organizer was delayed until 1929. It is not clear, however, why the report of the results was put off until 1931.

The occasion for the report was the meeting of the German Society of Zoologists that was held in May 1931 in Utrecht, at the invitation of the Dutch zoologists. Spemann gave a brief resume entitled "The behavior of organizers after destruction of their structure." He started as follows: "It is known that local actions of certain embryonic parts can invoke in other less differentiated regions specific formations, such as neural plate, lens, otocysts, or somites. One of the questions that arise immediately is the degree to which the struture of the implant, be it morphological or only chemical, is necessary for the induction of these structures in the tissues of the host" (1931b, p. 129). It should be noted that the emphasis is again on structure and that the meaning of this term is now stretched to include "chemical structure," though, of course, this is not what chemists would call "chemical structure."

Spemann used several methods to destroy the structure of the organizer: mincing, crushing, heating, freezing, and drying. Of these, only the first two were successful. In the first, pieces of the upper blastoporal lip were minced with glass needles and the hash was implanted into the blastocoele of an early gastrula. In three cases that were described, neural tubes were induced. The minced implants had undergone reorganization and had formed mesodermal structures. In the second experiment two kinds of implants were used: upper blastoporal lip from an early gastrula, or mesoderm and the ectoderm above it from a late gastrula. The tissue was pressed gently between two glass plates; care was taken to disarrange only the cytoplasm and to leave the nuclei intact. Using again the Einsteck-method, he obtained good neural inductions. In one particular case, a fairly complete secondary embryo with all axial organs and a tail bud was induced. The mesodermal structures were almost certainly derived from the reorganized implant. Spemann did not draw any specific conclusions from these experiments; obviously, he had not attained his objective, since the implants had been restructured to a considerable degree. At the end of the talk he acknowledged that a more radical procedure, including the destruction of the cell nuclei, would be required. But he himself did not follow up this suggestion. In fact, the experiments reported in Utrecht were his last personal contribution to analytical embryology, though he put his name on a later short communication from his laboratory that dealt with devitalized organizers. After his retirement in 1937, he did one more experiment: he transplanted gastrula ectoderm adjacent to the liver of an adult salamander to find out whether the fully differentiated tissue would have an influence on uncommitted embryonic tissue. The results were inconclusive. The manuscript that Spemann had prepared was published posthumously by O. Mangold (Spemann, 1942).

One cannot help asking: Why did Spemann go only halfway in devitalizing the organizer? I think what he did was in line with his scientific style and temperament. His reasoning was always very tight and precise. It motivated him to proceed in small steps and never to omit a step. This required self-discipline and patience. Others, with different temperament, would have taken the bull by the horns and killed the organizer outright. In this instance, Spemann's taking only half a step may seem pedantic. But it must be remembered that he had spent his formative years in close contact with Boveri, for whom the respective roles of nucleus and cytoplasm in heredity and development were a crucial issue.

Even though Spemann's experiments were no more than a first beginning, and not successful at that, he did break new ground. And his proposition to look at induction from a new angle was not lost on one young colleague in the audience, his former student J. Holtfreter.

The First Inductions by Devitalized Organizers

Spemann's experiment with minced organizer was the prelude to a set of more radical experiments that involved killing the organizer, performed in his laboratory and that of O. Mangold. They were published jointly in the December 1932 issue of *Naturwissenschaften* (the German equivalent of *Science* and *Nature*) under the title: "Experimental analysis of the inducing agents [Induktionsmittel] in embryonic development" (H. Bautzmann et al., 1932). This was a brief preliminary report, only three pages long and without illustrations. It was unusual in that each of the four coauthors, Bautzmann, Holtfreter, Spemann, and Mangold, reported his findings under a separate subheading.

Bautzmann, a former student of Spemann, was then on the staff of the Anatomical Institute of the University of Kiel in north Germany. He wrote on "Inductive capacity after killing by heat." He had done his experiments during a prolonged visit in Freiburg in 1927, probably at the instigation of Spemann, who was involved with his mincing and crushing experiments at the same time. Bautzmann had not published his results for a good reason: they were minimal and not very convincing. He had killed the organizer (upper blastoporal lip) or combined ectoderm and subjacent mesoderm of a late gastrula by submersion in hot water (60°C for 5 to 10 minutes). Using the Einsteck-method, he had found in several cases ectodermal thickenings that may or may not have been inductions, and in two of ninety-five cases a neural plate-like structure had developed that seemed to be in the process of upfolding to form a neural tube. These dubious findings attained some credibility when Holtfreter visited Bautzmann in Kiel in 1932 and showed him his positive cases obtained by the same method. Still reluctant, but apparently encouraged by Spemann, Bautzmann then decided to join the others. He said, "Having become acquainted with Holtfreter's results, I am now inclined to consider these two cases as induced neural plates" (H. Bautzmann et al., 1932, p. 972).

Holtfreter had moved to Berlin-Dahlem in 1928, succeeding me as an assistant (research associate) in O. Mangold's Division of Experimental Embryology at the Kaiser Wilhelm Institute for Biology. As was mentioned, in 1931 he had devised a culture medium, a modified Ringer's solution, later known as Holtfreter's solution, which, together wtih the application of strict antiseptic methods, reduced the mortality of operated embryos very substantially. His remarkable success of 1932 would not have been possible without these improvements. His contribution to the joint publication was experiments performed in 1932. They had been inspired by Spemann's presentation in Utrecht. The title of his section was "Inductive capacity of dried, heated, and frozen embryonic parts." To test the inductive capacity of the dead tissues, he had designed two new methods. In one, the tissue was dried on a petri dish and a piece of gastrula ectoderm was placed on top of it after culture medium had been added. In forty positive cases, brain parts and neural tubes were induced. To his surprise, he found that dried gastrula ectoderm and embryonic intestinal cells would also induce neural structures in the gastrula ectoderm. In all these experiments, a 1-to 2-day contact with the substrate was required. The discovery that tissues that had no inductive capacity when alive acquired it after they had been devitalized was of signal importance; it set the analysis of embryonic induction on a new course. From now on, induction by atypical inductors became a major issue.

In a second experiment, the dead tissues were placed between two large pieces of gastrula ectoderm which healed together and formed a flat vesicle around the implant. This method became known later as the "sandwich experiment." In thirty cases, neural structures were induced in the wrapping. Finally, in sixty cases, the Einsteck-method was used; it gave the most elaborate inductions

> These attained enormous size and resulted in very complex structures such as brains with eyes, noses, and typical balancers which were histologically indistinguishable from inductions by living inducers. Again, dead upper blastoporal lip, different parts of the neural plate with subjacent mesoderm, killed endoderm and prospective epidermis, and even fragments of a hard-boiled uncleaved egg were equally effective. . . . The kind of killing—drying at 60°C, heating to 100°C, freezing—did not seem to have an essential influence on the kind of induction. (H. Bautzmann et al., 1932, p. 973)

The title of Spemann's section was "Inductive capacity after killing by alcohol." This was a report on a single case:

> To the fine success of Holtfreter in an experiment in which my own endeavors had been unsuccessful, I can at least add a small positive result obtained in my laboratory. Miss E. Wehmeier, who is involved with experiments on the behavior of dried and frozen embryonic parts in epidermal vesicles, had the idea of testing the inductive capacity of embryonic parts which had been placed in alcohol for some time. She obtained the induction of a fine neural plate by a dorsal plate [ectoderm and subjacent mesoderm] of a late gastrula which had been lying in 96% alcohol for 3½ minutes. (H. Bautzmann et al., 1932, p. 973)

The fact that only a single case was reported makes one wonder whether this was actually a planned experiment. It was rumored that Wehmeier had immersed the organizer in alcohol by accident, had then rescued it and used it as an implant. Spemann drew the obvious conclusion from this and the other experiments that the inducing agent is chemical in nature and that it is not soluble in water or alcohol. Then he pointed out a number of problems which now became acute, as for instance the question of whether the hypothetical agent is taken up by the responding tissue.

O. Mangold, who had witnessed Holtfreter's successful experiments in his laboratory, went to work immediately on an idea of his own. His contribution was entitled "Is the inductive agent diffusible?" He tried to obtain inductions indirectly, by transmitting the putative agent by way of an inert carrier. He dried a neural plate on agar, allowed 2 to 26 hours for the postulated agent to be absorbed, then removed the dry material carefully, cut out the agar piece, and inserted it in the blastocoele of a gastrula. A number of implants were extruded. In a few cases, the ectoderm showed irregular thickenings, and in a single case, a small induced neural tube was found. Mangold did not pursue his experiments any further. But many years later the Finnish experimental embryologist S. Toivonen looked into the matter: "To my knowledge, nobody has been able to repeat his [Mangold's] experiment with positive results. I myself have tried to do so, but with negative results. I am afraid that Mangold's positive result must have been caused by contamination; some cell detritus might have remained on the agar and caused the reaction" (Toivonen, 1978, p. 142).

I suppose I was not the only reader of this article to be puzzled by the contrast between the 130 manifestly positive cases of Holtfreter and the four weak cases (one positive, two dubious, one spurious) by the other three coauthors. One wonders how this publication originated. Fortunately, I can provide the testimony of two of the contributors; it throws some light on the matter. Several years ago, Holtfreter wrote an autobiographical essay at the request of the Japanese developmental biologist, K. Ishihara. It was translated into Japanese and published with a foreword by Ishihara in the Japanese journal *Shizen (Nature)* in 1982 (J. Holtfreter, 1982). It is entitled "Recollections of an embryologist—in search of the organizer." With the consent of my friend Holtfreter I quote the pertinent passages from his English version:

> In 1931, at a metting of the German Zoological Society which was held in the beautiful city of Utrecht, Spemann lengthily reported on his negative results with crushed blastoporal lips. I attended the meeting and, like so many others in the audience, was rather disappointed. I thought: Is this the end of all endeavors to analyze the mechanisms of neural induction? I could not believe it. . . . Here was a problem of fundamental importance that cried out for a resolution. I decided to do something about it myself. Returned to Dahlem, I immediately began to work on this project.

He then describes the three types of experiments which I have discussed above, and continues:

> Now the question arose what to do with these revolutionary data? I had not behaved like a faithful disciple of Spemann; I had failed to tell the master what I had been doing lately, and done it without his explicit consent. Now Mangold, who so far had been a benevolent observer of my doings, urged me to communicate these findings to Spemann. This I did with some trepidation. Spemann applauded but did not actually congratulate me. He pointed out in his letter that some three years ago, Hermann Bautzmann had been given the green light to work on heat-killed inductors. Although Bautzmann had as yet not published anything about this work, I should consider his right of priority and should not publish my findings until I had discussed them with Bautzmann. So I travelled to Kiel to see my friend Bautzmann, carrying along with me some of my prettiest microscopic slides. When he examined these slides, Bautzmann wondered how I had been able to raise my experimental material up into highly advanced stages of differentiation. So I told him about my new culture method. He however had been plagued by the old trouble: His operated embryos had usually perished before they had reached the tail bud stage. Therefore, as to the effect of heat-killed organizer, his results were not very impressive. There were only a few cases which, at best, showed the induction of small lumps of barely differentiated neural tissue but there were none of the massive brain formations I had obtained. It was a friendly meeting. We agreed to publish our findings jointly.
> In the meantime, Mangold had been busy. My experiment had indicated that the neuralizing agent is diffusible. Mangold now had the idea to let the hypothetical agent, derived from an isolated piece of inductor, diffuse into a piece of agar and then implant this agar into the early gastrula. He did not have the time to make more than a few preliminary experiments of this kind.

Then Spemann, whom I had informed about my agreement with Bautzmann, expressed his wish to join us and to report on some pertinent findings of a young student of his, Else Wehmeier.

Thus it came to pass that the four-men communication of the inductive mechanism in embryonic development saw the light of publication (*Naturwissenschaften*, 1932)

Spemann's comment on the same publication is contained in a letter to me from Freiburg, dated December 25, 1932 (I was then a Rockefeller Fellow in the zoology department of the University of Chicago): "In the December issue of *Naturwissenschaften* you will find an article by four authors, Bautzmann, Mangold, Holtfreter, and myself. Another dream of mine has been fulfilled, and from a threatening collision [came] an amiable cooperation. More importantly, a new beginning of very promising investigations has been made." Apparently, Spemann's reference to a potential "collision" in the matter of priority relates to difficulties that might have arisen if Holtfreter had published without Bautzmann's consent. This was avoided. It is clear, however, that the joint publication was orchestrated by Spemann and that his intention was to give his friends H. Bautzmann and O. Mangold and his student Wehmeier a chance to share the priority of this important discovery with Holtfreter. But the participation of the minor authors in this episode was soon forgotten, and history has settled the issue, assigning the priority of the discovery jointly to Spemann, who had the idea, and to Holtfreter, who executed the experiment with striking success.

The experiment was a turning point in more than one way. First and foremost, it gave the analysis of induction an entirely new direction: it shifted the emphasis from the inductor as a tissue to a possible chemical inducing agent contained in it and, inevitably, to the properties of the reacting system that enabled it to respond to the activating agent with complex differentiations. The discovery of "abnormal" inductors raised hopes that the analysis of induction could be expanded in new ways, and this hope was soon fulfilled when a large assortment of living and dead tissues was found to induce not only neural differentiation but other embryonic structures as well.

Second, this was the beginning of a new era in a more general sense. What so far had been essentially a family enterprise of Spemann and O. Mangold and their associates and students, now, almost overnight, became a cosmopolitan venture. Intense activity in search of inductive agents began in a number of European laboratories and in Japan. At the same time, the leadership, at least in the area of experimental embryology that was concerned primarily with induction, passed from Spemann to a younger generation with a different outlook and with new methodological resources. Spemann became an interested observer. On the eve of the fateful juncture, in the fall of 1931, he had received the invitation to give the Silliman Lectures at Yale University in 1933. He took this opportunity to synthesize the old and new ideas in a grand overview. The lectures were published in book form. The German edition was published in 1936; it was the result of four years of arduous work which took all his energies, as he stated—and bemoaned—in several letters to me. The English translation was published in 1938, with no additions but with the omission of a chapter on chick, echinoderm, and insect embryos.

CHAPTER 6

New Approaches to Problems of Determination and Induction: The Ascendance of J. Holtfreter (1931–1934)

The communication of 1932 set in motion a sequence of activities that followed two distinctly different paths. One group of experimental embryologists joined forces with biochemists in the search for the chemical identity of inductors, and specifically of the neural inductor. They were guided by the observation that the preliminary experiments with killed embryonic tissues had resulted invariably in the induction of neural plates. Spemann's student, Else Wehmeier, collaborated with the Freiburg chemist, F. Gottwald Fischer. In Cambridge, England, intensive activities were masterminded by the historian and practitioner of chemical embryology, Joseph Needham. He was joined by the experimental embryologist and developmental geneticist, Conrad Waddington. Soon, other laboratories took up the biochemical analysis of inductors, among them those of Jean Brachet in Brussels, and later those of Sulo Toivonen in Helsinki and Tuneo Yamada in Nagoya.

The reappearance of Holtfreter on the scene marked the beginning of a quite different venture. While the biochemical approach favored teamwork, Holtfreter, single-handedly, launched a vigorous attack on problems of induction in which, in addition to the Einsteck-method, he used novel methods of his own, some of which have been mentioned. More importantly, he brought a new perspective to the treatment of old problems. Gradually, he transformed Spemann's vision of development, which had a decidedly holistic tenor, and moved toward a more reductionist viewpoint. Historically, the change was marked not by a revolutionary breakthrough but by a gradual shift of emphasis. At the beginning, Holtfreter followed closely in the footsteps of his mentor. He acknowledged this in the introduction of his publication on the isolation of gastrula parts *in vivo,* which I shall discuss presently: "Among the questions which can be resolved in this way, I took up first the question of the potentialities of the earliest embryonic stages; this came nearest to my interests as a former student of Spemann" (J. Holtfreter, 1929b, p. 422). And I have mentioned already that another project of his, the inductions by killed embryonic parts, had been inspired by Spemann's talk in Utrecht. A decade later, he began to look into the activities of individual embryonic cells and cell assemblies. This topic then acquired a momentum of its own, and he is now remembered best as the discoverer of tissue affinity and disaffinity (J. Holtfreter, 1939).

If I pay much attention to the contributions of Holtfreter, I do so first and foremost because I consider his work of the late 1920s and 1930s as of signal importance in a period of transition in experimental embryology, but also as a testimony to a lifelong friendship, and an acknowledgment of his influence on the shaping of my own ideas about basic issues in embryonic development. Our friendship started in 1920, when we worked side by side in Spemann's laboratory in Freiburg, getting our first exposure to the amphibian embryos as objects of scientific studies—we both had collected salamander and frog eggs since our early youth. We thus shared the trials and tribulations that are the lot of Ph.D. candidates. But at that time we did not communicate much, partly because we had different working habits— Hannes being a night owl while I preferred the daylight—and partly because we moved in different student circles (see the appendix). After we had earned our Ph.D.s, we spent a few months at the Stazione Zoologica in Naples to broaden our zoological horizon. We roamed the countryside and the beautiful islands of Ischia and Procida. I have memories of volcanoes of all sizes and of huge oranges in the vast orange groves. Soon, Hannes found the bucolic atmosphere of Ischia more to his liking than the laboratory. He moved in with a baker's family in a small village on Ischia and indulged in his favorite pastime of painting and drawing. He soon became known among the villagers as "Giovanni il pittore" and he painted their patron saint, St. Angelo, for their village church. In the meantime, I studied dutifully the rich harvest of marine animals which the collecting boat of the Stazione brought home daily at 10 o'clock in the morning. I was then on the dock and took possession of that part of the catch that was not claimed by the international group of distinguished researchers. I was rewarded for my labor by the exquisite beauty of the Mediterranean marine fauna.

We lost contact for several years. I took a brief fling at comparative physiology at the laboratory of Professor Alfred Kühn in Göttingen, the universal genius among the leading German zoologists. Apart from a widely used textbook of general biology, he wrote books on comparative physiology, experimental embryology, and genetics. His expertise and his art of articulating his knowledge in lectures and books were exceptional. He had just introduced the spectroscope to the study of color vision in animals and I used this tool to study the color vision of fishes (Hamburger, 1926). I then returned to experimental embryology and to amphibians as an assistant (research associate) in O. Mangold's division at the Kaiser Wilhelm Institute for Biology in Berlin-Dahlem. Meanwhile, Holtfreter had been struck by Wanderlust. For three years he gratified his love of adventure, traveling to remote countries, not as a tourist but with a knapsack, living among the natives, exploring their customs and occasionally earning his meals and lodging by sketching the portraits of his hosts; science was forgotten! When the time came to settle down, fortunately for science, he accepted in 1928 the position in O. Mangold's laboratory which had been vacated by my return to Freiburg. His five years in Berlin-Dahlem were among the most productive of his life. In 1933 he moved to Munich as an associate professor in the Department of Zoology, whose director was Professor Karl von Frisch, the discoverer of the language of the bees. The Munich period was interrupted in 1935 by a one-year Rockefeller Fellowship which he spent at the laboratory of Ross Harrison at Yale. Another grant enabled him to return by way of Hawaii, the Far East, and the South Sea Islands. His most memorable experience there was his stay, for several months, on the island of Bali. He felt very congenial with the natives, who are renowned for their sense of beauty and their skills in the arts, music, and dance. He brought home dozens of black and white drawings of landscapes, native scenes, and portraits of the islanders done in his favorite medium, the scratchboard technique.

He left Nazi Germany voluntarily in 1938 and found refuge in Cambridge, England, where his friends Needham and Waddington provided him with laboratory facilities. When war broke out a year later, he was declared an enemy alien, and in 1940 he was sent to Canada where he spent two years in a labor camp. In 1942 his Canadian and American friends succeeded in obtaining his release, and he worked then as a guest in the Zoological Laboratory of McGill University in Montreal: its chairman, Dr. N. Berrill, became his friend. In 1946 he accepted a professorship in the Biology department of the University of Rochester and he has resided there ever since. Our contacts and visits were resumed after his arrival in the United States. Our friendship survived a severe test when we agreed to collaborate on the chapter on Experimental Embryology of Amphibians in the book Analysis of Development, *which was published in 1955. It was not easy to reconcile our different styles of reasoning and writing. An incessant exchange of new drafts, revisions, and shortenings, accompanied by comments that ranged from the acrimonious to the hilarious, shuttled back and forth between Rochester and St. Louis.*

I return now to Holtfreter's beginnings in Berlin in 1928. The years away from science had some very positive effects on his new adventure. Although he started again where he had left off in Freiburg in 1925 and, for a while, followed in the tradition of Spemann, as I have mentioned earlier, he had acquired an independence of judgment and a nonconformist stance that enabled him to move faster and farther beyond Spemann's orbit than others, such as O. Mangold or H. Bautzmann, who hardly ever transgressed it. He simply was bolder and more inventive and willing to take risks. Furthermore, he outdid everybody else in the sheer output of work. In the years of freedom from the bondage of science he seemed to have stored boundless energies that were now released. His endurance seemed to know no limits. He immediately tackled three or four projects simultaneously and did many hundred experiments for each. At first he had only part-time help and did much of the sectioning and staining of preparations for the microscopic observations himself. His first full-time technician happened to be an acquaintance of my family; she told us toward the end of 1932 that he had done over 4,000 exper-

Fig. 6–1. Johannes Holtfreter (1971).

iments in that year, that he was writing half a dozen papers, and that he worked till 2 or 3 o'clock each morning. In fact, he issued seven publications in 1933, three of which were substantial reports on major projects. In 1934, three major and several minor publications appeared, and his production continued at that pace until 1939. Work was interrupted only by the trips to the United States and around the world in 1935–36 which I have mentioned above. (We others managed at best a few hundred operations in a breeding season, but it should be added that not all of Holtfreter's experiments required difficult transplantations.) His artistic talent was also put to work. Of the hundreds of drawings he made of experimental embryos, quite a few found their way into his publications. He gratefully acknowledged the generous support that O. Mangold gave him, but he confided to me once that he did not always tell Mangold "everything."

In Vivo Isolation Experiments and a New Culture Medium

Holtfreter inherited from Spemann's analysis of development two central themes: the state of determination of different embryonic parts at different stages of development, and the interactions between the parts by which their programming was achieved. While transplantation was the method of choice for tests of interactions and inductions, it was evident that this method could not give unequivocal results concerning self-differentiation and that a method of complete isolation of parts was required. Efforts to grow embryonic materials in tissue culture, following the pioneer experiment of Harrison, had been unsuccessful in the hands of Spemann, Geinitz, and O. Mangold. Holtfreter, who had tried this method from 1928 on, had also failed so far. While he continued his efforts, he fell back on an *in vivo* isolation method which he had worked out for his Ph.D. thesis. Spemann had assigned to him the task of investigating the determination process of the liver and its accessories, the pancreas and gall bladder. These organs are endodermal derivatives; their embryonic primordia are notoriously difficult to handle because they are composed of heavily yolk-laden cells that cannot be implanted on the surface of embryos, as could ectodermal or mesodermal parts; and the Einsteck-method was not feasible either. Hence, Holtfreter had to devise a suitable isolation method before he could start the analysis. He did find an implantation site that seemed to fulfill all prerequisites. He used the tail bud stages of the embryos of the common toad *(Bombinator pachypus);* he made a small slit in the belly region, removed a small amount of yolky cells, and pushed the transplant rather deeply into the slit. The wound healed over and the transplants were found later in the coelom (body cavity) of the tadpoles where they floated freely or were attached to the coelomic wall. The anlagen of liver, pancreas, and gall bladder differentiated normally to advanced stages in this milieu. The actual results of the investigation are of no particular interest at this point. It turned out that the precursor material of the three structures is already programmed at the earliest stage at which it can be localized, that is, in the late gastrula (J. Holtfreter, 1925). To the student of the determination process, this was not a particularly illuminating finding. But the new method of growing embryonic parts in isolation had its merits.

One of the first projects Holtfreter undertook when he opened his workshop in Dahlem was to use his *in vivo* method for a comprehensive analysis of the state of determination of all parts of the gastrula. He divided the early gastrula into about twenty subregions and implanted as many parts of a single gastrula as possible into a set of host embryos. At that time the fate maps of the gastrula which

Vogt had designed on the basis of his vital-staining experiments had become available; they made it possible to find out whether or not a given implant would differentiate according to its normal fate. Most of the experiments were done xenoplastically with salamander embryos as donors and frog or toad embryos as hosts.

The experiment was a success. In over 2,000 implantations, Holtfreter established that the coelomic milieu is suitable for the full differentiation of all implanted gastrula parts: the xenoplastic exchange gave the same good results as the homeoplastic transfer. The differentiations proceeded to advanced stages, such as cartilage, muscle, notochord, and intestine. Meticulous studies were made of the growth rate, blood supply, and other aspects of their development, but these do not concern us here. As a general rule, all mesodermal and endodermal derivatives differentiated according to their normal fate (J. Holtfreter, 1929b). The ectodermal implants received special attention; their behavior was reported separately at two meetings of the German Zoological Society in 1929 (J. Holtfreter, 1929a) and in 1931 (J. Holtfreter, 1931b). I shall consider only the second communication, which supersedes the first. The results, based on a large amount of material, were puzzling, indeed. As it turned out, this was the beginning of the unraveling of the story of *in vivo* implantation. Holtfreter found that prospective epidermis and prospective neural plate differentiated to ciliated epidermis and to neural tissue at the same frequency. Quite often, both types of tissue were found side by side in the same implant. These results were disturbing and ambiguous; they gave no clear evidence of the state of determination of the gastrula ectoderm. The situation became utterly confusing when other findings of *in vivo* implantations became known. At about the time when Holtfreter had devised the method of implantation to the coelom, the German experimental embryologist B. Dürken had designed a similar experiment. He had grown tissues in the orbit of tadpoles after removal of the eye (Dürken, 1926). Dürken's student W. Kusche had implanted gastrula parts into the orbit and found that not only prospective mesoderm but also ectoderm proceeded to differentiate to notochord and muscle (Kusche, 1929). Simultaneously and independently, Bautzmann had had the same idea. The implantation of gastrula ectoderm into the empty orbit of salamander larvae resulted likewise in the differentiation of notochord and muscu-Bautzmann, 1929). One was thus confronted with the paradox that the same material, gastrula ectoderm, formed the two normal ectodermal derivatives in the coelom but exclusively mesodermal derivatives in the orbit.

It is very revealing to see how two personalities as different as Bautzmann and Holtfreter handled the situation. Bautzmann persuaded himself that the orbit was a neutral medium. "It seems to me improbable that in the method of culturing in the orbit [extrinsic] factors came into play which could influence the result in a specific way. . . . As far as we know, hormones in a wider sense can only enhance or inhibit preformed mechanisms. . . . One could imagine hormones in the orbit: their effects would only be to activate differentiations, such as the development of muscle and notochord, but not to create new ones" (H. Bautzmann, 1929, p. 823). The presumption that the milieu of the orbit is free of any determinative influence led Bautzmann to the conclusion that the gastrula ectoderm had the inherent capacity for "self-differentiation" to notochord and muscle, before it was transplanted—a rather unorthodox view, considering the normal path of differentiation of the ectoderm, which, incidentally, seemed to have been followed faithfully in Holtfreter's coelomic grafts. To conceptualize his view, Bautzmann coined the term "bedeutungsfremde Selbstdifferenzierung," which can be translated as "self-differentiation not conforming to the normal fate." The term seemed to be a contradiction in itself, but Bautzmann did not see it this way. He had a penchant for theoretical speculations and elaborated the general idea of multiple self-differentiation capacities in a fanciful hypothesis which has a distinctly preformistic flavor. He imagined that the blastula and early gastrula are composed of an agglomerate of mosaic stones or building blocks, each capable of self-differentiation to a single specific structure (muscle, kidney, etc.). Progressive differentiation would then consist of regional suppression of the inappropriate units, and terminal differentiation of a region would be governed by those units which had not been eliminated. But if the autonomous fate of gastrula ectoderm is notochord and muscle, how about its fate in normal development? Bautzmann is not very explicit about this point, but he hints at what would seem to be the logical interpretation according to his model: that the organizer would suppress in the neural plate region the notochord and muscle units, so that the neural units could prevail—an implausible reinterpretation of organizer action. Altogether, Bautzmann's ideas were never elaborated; the promised detailed account of his experiments never saw the light of day. No wonder that this figment of his imagination was soon forgotten. I have revived it only to contrast this kind of speculation with the theoretical conceptions of Spemann and Holtfreter which were always in tune with the embryo.

Holtfreter, the pragmatist, soon became suspicious of the trustworthiness of the *in vivo* method as a test for self-differentiation and he redoubled his efforts to devise an inorganic culture medium. Finally, after several years of trials, he met with success in the spring of 1930. I shall deal with his major publication of 1931 presently, but first I shall anticipate the results of the explantation of gastrula ectoderm, which settled the controversial issue of its state of determination once and for all. Both prospective ectoderm and prospective neural plate uniformly differentiated *in vitro* to epithelial cells; neural differentiation never occurred (J. Holtfreter, 1931b). This was convincing evidence that the *in vivo* method was flawed and that the neuralization of ectoderm in the coelom as well as its mesodermization in the orbit was due to uncontrollable agents produced by the host. While Bautzmann's theoretical construct fell apart, Holtfreter simply shelved his 2,000 *in vivo* experiments and the labor invested in them, and repeated and actually expanded the potency tests of early stages with the *in vitro* method.

From the beginning, Holtfreter had set his goals higher than just to settle the old dispute of the state of determination of the gastrula ectoderm:

> The need for an isolation method which permits the culture of the material entirely outside the organism, that is, *in vitro,* becomes more and more urgent. . . . I was guided by the special desire, inspired by the problematics of determination, to exclude all possible unknown factors which might have a specific influence on the differentiation of the explants; they can never be excluded

with certainty when one uses living or organic media. It is also of importance, on general methodological grounds, to devise a culture medium which would remove from the whims of chance not only the maintenance of embryonic fragments but that of whole experimental embryos and to guarantee their longer survival. (J. Holtfreter, 1931a, pp. 406–407)

The inorganic salt solution that fulfilled these requirements was a hypertonic Ringer's solution including a buffer. The original solution was improved later by the addition of magnesium ions. Holtfreter also emphasized the necessity for strict sterilization, which had not always been taken seriously. Holtfreter's solution, which the originator referred to as the "standard solution," immediately became indispensable in all laboratories that used amphibian embryos. It saved endless hours of labor and, most importantly, it made possible the solution of problems that required long-term culturing periods and complete tissue differentiation. But the comprehensive account of the new method is not concerned with analytical problems of determination. It is devoted almost entirely to technical details such as rates of development and survival times; there is no need to discuss them in detail. Only one particular observation deserves mentioning, because it was exploited later for analytical purposes. If all protective membranes are removed from an early gastrula, including the innermost vitelline membrane, which holds the soft egg mass together, then the embryo flattens and the hypertonicity of the culture medium interferes with the gastrulation movements. The ventral hemisphere evaginates instead of invaginating, that is, it turns outward rather than inward. The process is referred to as "exogastrulation." Since the empty ectodermal vesicle never showed any sign of neural differentiation, it provided another piece of evidence that neuralization requires an inductive stimulus.

Region-Specific Inductions in Older Embryos

In retrospect, the protracted debate on the issue of labile predetermination of gastrula ectoderm seems somewhat overdone. Holtfreter took another try at it; but what started as the rebuttal of another ill-conceived claim of Goerttler led to a deeper understanding of induction, thanks to Holtfreter's broad perspective. Goerttler, whose previous encounter with Holtfreter has been dealt with above, shared with Vogt, his mentor, two theoretical tenets: a belief in the early predetermination of embryonic primordia, and in the overriding importance of morphogenetic movements as integral parts of the determination process. While Vogt took a cautious stand, particularly on the second point, his disciple was more doctrinaire and eager to prove his point. To understand the intent of his experiment, I have to go back to his earlier work. While Vogt was engaged in his vital-staining experiments dealing with fate maps and gastrulation movements, he delegated to Goerttler the task of studying with the same method the shifts of material that transform the sickle-shaped prospective neural area of the early gastrula into the pear-shaped neural plate, and the subsequent upfolding of the margins of the neural plate to form the neural tube. Goerttler gave an excellent account of the flanking movements that accomplish the first phase and of the upfolding in the second phase. He then turned to analysis and designed an experiment which he thought could prove that the morphogenetic movements are a necessary prerequisite for neural differentiation. Focusing on the second phase, he tried to control the direction of stretching of a transplant by placing it close to the rising folds of a neurula. He implanted a piece of ectoderm from the prospective neural plate region near the outer margin of the neural plate of an older host. He argued that if the transplant were inserted in a "favorable" position, so that its inherent stretching tendencies were concordant with those of the upfolding margin of the host neural plate, then it should differentiate to neural plate; if it were placed in an "unfavorable" position, with its inherent direction of stretching at a right angle to that of the folds of the host, then neural differentiation should not occur, and epidermis should differentiate. His experiments seemed to support his hypothesis (Goerttler, 1927).

Holtfreter had good reasons to be skeptical, considering the new results of the *in vitro* experiment. He repeated Goerttler's experiment on a much larger scale. The title of his publication tells of his results in a nutshell: "Not typical morphogenetic movements but inductions cause neural differentiation of gastrula ectoderm" (J. Holtfreter, 1933a). Prospective neural ectoderm became neuralized in 80% of the cases, irrespective of the direction of its implantation. Furthermore, prospective epidermis of the early gastrula likewise became neuralized, and again the direction of its implantation was irrelevant. To exclude "dynamic" factors altogether, Holtfreter repeated the experiment using older hosts in which the folding movements had come to rest and the neural tube was closed. The results were the same. He came to the conclusion that the only explanation for neural plate and tube formation by the transplants was induction by the subjacent somites. "In all cases which showed neural differentiation in the trunk region, only a determinative influence of the mesoderm can come into consideration" (J. Holtfreter, 1933a, p. 613). According to Holtfreter, most of Goerttler's cases had been sacrificed too early, before they could be identified as distinctly neural or epidermal; in other instances, the transplants had been placed too far ventrally, out of the reach of mesodermal inductors.

Holtfreter realized that the experiment, which had been undertaken originally as a test of an improbable claim, based on poor experimental evidence, had much greater potential. He took his cue from the fact that the somites had retained their capacity for neural induction long after they had participated in the primary induction of the neural plate as part of the organizer. Two questions of general interest could now be addressed: (1) For how long does the inductive capacity persist? (2) Do different levels along the rostrocaudal axis of the host embryo induce region-specific structures? Holtfreter undertook an extensive series of experiments to answer these questions. The first was addressed by using host embryos of different stages, from neurulae with rising folds to advanced tail bud stages. The operations were done xenoplastically between salamander and axolotl *(Ambystoma mexicanum)* embryos, so that host and induced tissues could be distinguished on microscopic preparations. It was found that the somites of early tail bud stages could still induce complex structures, but that their inductive capacity vanishes at later stages.

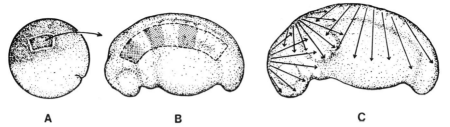

Fig. 6-2. a,b. Transplantation of prospective ectoderm of the early gastrula to three regions on the flank of tail bud stage. c. Diagram of three induction fields: anterior head inductions, posterior head inductions, trunk inductions. From J. Holtfreter, 1933c.

To test for regional induction-specificity, the same series of stages was used, and three different implantation sites were chosen: the posterior head region, the gill region, and the trunk region (Fig. 6-2). As usual, the transplants were taken from early gastrulae. A large number of complex inductions was obtained, representing the entire spectrum of embryonic structures such as well-organized brain parts (Fig. 6-3) and fully differentiated notochord and musculature (Fig. 6-4). The inductions were region-specific in that they showed clear relations to the host levels at which they were produced. The statistical analysis of the frequency of particular inductions at different host levels indicated that head structures were

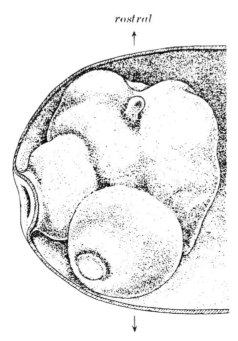

rostral

Fig. 6-3. Induction of symmetrical forebrain with epiphysis, two eyes, and nose in the posterior head region. Operation as in Fig. 6-2a,b. From J. Holtfreter, 1933b.

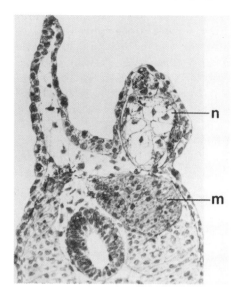

Fig. 6–4. Induction of notochord (n), somitic musculature (m), and tail fin (left) in the trunk region. Operation as in Fig. 6–2a,b. From J. Holtfreter, 1933b.

induced most frequently at the head level and trunk structures at the trunk level. However, there was a considerable spread to adjacent host regions, though the highest percentage of an induced structure was found at the corresponding host level. For example, of the ear vesicle inductions ($n = 192$), over 70% were located at the ear level of the host, 15% at the level of the gills, and 4% at the trunk level. The notion of *gradient fields of induction* suggests itself. Holtfreter comments: "These findings lead to the assumption that the inductive effect is not tied strictly to the immediate neighborhood of the inductor of a particular tissue but that it spreads frequently beyond it with a decrease of intensity towards the periphery, after the fashion of a 'determination field.' Such fields overlap, whereby perhaps mutual inhibitions or intensification can be expected" (J. Holtfreter, 1933c, p. 424). I have found the term "determination field" rather vague and have used the term "induction field" instead. Holtfreter's diagram (Fig. 6–2c) illustrates the extent of three overlapping fields: the forebrain-eye field, the hindbrain-otocyst field, and the spinal cord-pronephros field.[1] One feature of the sketch is somewhat misleading. The direction of the arrows is not meant to indicate the spreading of hypothetical inductive agents from dorsal to ventral. Holtfreter states repeatedly that inductions occur locally by contact with the host tissues adjacent to them. However, he was unable to identify specific inductors of specific structures. This is perhaps the weakest point of the experiment and probably the reason why it was not pursued any further.

We may ask: What is actually novel about the experiment? Regional specificity of inductions had been discovered by Spemann (1931a) when he found the difference between head and trunk organizers. What Holtfreter has added is the demonstration of *gradient fields* of induction. What enabled him to go beyond

[1]The term "field" is used here in a broader sense than in all other usages of this term in this book.

Spemann was the fact that he dealt with relatively small, distinct structural units rather than with whole body regions. Inductions by older tissues were not a novel phenomenon either. O. Mangold (1929) had shown that brain fragments of swimming larvae can induce neural plates in Einsteck-experiments. But in these and in the earlier experiments of Mangold and Spemann (1927) only neural structures had been induced (homeogenetic inductions). In contrast, in Holtfreter's experiments, the different body regions of the host embryos had induced a great variety of both ectodermal and mesodermal derivatives, homeo- and heterogenetically. All this meant a considerable extension of our insight into the nature of embryonic induction: both the *in vivo* experiments and the transplantations of ectoderm onto older tissues bring into focus the enormous range of reaction potencies (competence) of gastrula ectoderm, as well as the complexity of abnormal inductors. Both aspects constitute the foundation of the large array of experiments with which we shall deal in the following chapters.

One particular facet of abnormal inductions that has not been mentioned as yet is of special interest. It is directly related to the age discrepancy between the inductors and the gastrula ectoderm transplants. Many inductions are highly complex and well organized within themselves (Fig. 6–3). Yet they are never integrated harmoniously in any part of the host embryo. The induced brain parts remain separate from the host brain; induced muscle does not fuse with host muscle; the induced kidney tubules are not connected with those of the host, although they may be adjacent to them. All inductions in the older embryos are supernumerary structures, and in this respect abnormal. One can understand this if one realizes that at the stages when the inductions occurred the host embryos were already complete self-contained units and "the size and pattern of the [inducing] anlagen of the host were already determined" (J. Holtfreter, 1933b, p. 749). As a result, the host structures had lost the capacity for assimilation of the extraneous material.

It is very illuminating to contrast this aspect of induction by older embryos with the inductions that occur in normal development when no such age difference between inductor and reacting tissue exists. Take the two classical cases: lens induction and induction of the neural plate. One comes to realize that these normal inductions fulfill two functions: they initiate new structural differentiations, and at the same time the inductor and the induced structure become integrated as parts of a higher organizational unit: the eye in the first instance and the axial organs in the second. A precondition for this double function is that the inductor and the reacting tissue are close neighbors and of the same age. It is this double role which gives embryonic induction its unique place in vertebrate development and distinguishes it from other interactions such as trophic interactions in the developing nervous system, growth factors, and hormones. The latter can operate by remote control and by signals because they are not required to fulfill a direct integrative function. Implicitly, the two aspects of induction had been recognized before, but by dissociating them, the Holtfreter experiment brought them into bold relief.

Inductions by older tissues are atypical in other respects. The small patches of ectoderm are stimulated by the subjacent somites to produce massive structures which include not only the normal ectodermal derivatives—brain, spinal cord, and epidermis—but also other structures, such as muscle, notochord, kidney, which are not in its normal repertory. In previous experiments, mesodermization of gastrula ectoderm had occurred only by assimilative induction, as in the experiment of O. Mangold (1923) when he implanted prospective ectoderm in the upper blastoporal lip. In Holtfreter's experiment, mesodermization was brought about in a very different way. The inner layer of the ectoderm grew inward, proliferated profusely, and differentiated to form structures that are normally derived from the mesodermal sheet. The details of these processes, which depart radically from the normal developmental pathway, are not clearly understood, and I shall not pursue this issue at this point. But the two major aspects of Holtfreter's experiment—demonstrations of the inductive capacity of older tissues and of the almost unlimited range of structures which the gastrula ectoderm can produce when stimulated properly—were on his mind when he undertook tests of the inductive capacity of living and dead embryonic and adult tissues, in a variety of later experiments.

Exogastrulation

The formation of exogastrulae was a fringe benefit of Holtfreter's design of a culture medium for gastrula parts. As I have mentioned, the gastrulation process was severely disturbed when whole embryos were exposed to the hypertonic Holtfreter's solution. In the majority of cases, invagination began in a normal fashion; but sooner or later it reversed its direction, resulting in partial exogastrulae. In its extreme form, the reversal began almost immediately after the onset of the gastrulation process; this led to total exogastrulation. The entire mesoderm and endoderm moved outward, leaving an empty bag of ectoderm behind. The two parts remained connected by a narrow stalk (Figs. 6-5, 6-6. For a comparison with normal gastrulation, see Figs. 2-6, 2-7).

Exogastrulation had been known since the 1890s. Herbst (1893) had made the discovery that sea urchin embryos would exogastrulate if lithium salts were added to the sea water. This classical experiment introduced the phenomenon of exogastrulation to experimental embryology. Soon it was repeated on amphibian embryos by several authors, including Vogt (1922). But Holtfreter was the first to obtain *total* exogastrulation in amphibians by a special technique. In previous experiments, partial exogastrulation had been obtained by removing the jelly membranes from the eggs and then placing the embryos in a hypertonic salt solution. Holtfreter removed, in addition, the innermost, very thin, transparent vitelline membrane that holds the soft egg mass together. This must be done several hours before gastrulation begins. This procedure was not new; it had been done routinely for preparing early embryos for transplantation and other experiments. It was the combination of this expedient with the choice of suitable egg material that enabled him to obtain total exogastrulation. The physical consistency of amphibian eggs varies considerably with the species. For instance, the eggs of *Triturus taeniatus* and *T. alpestris* are more firm than those of *T. cristatus*. Hence the latter are preferred as donors and the former as hosts in transplantation exper-

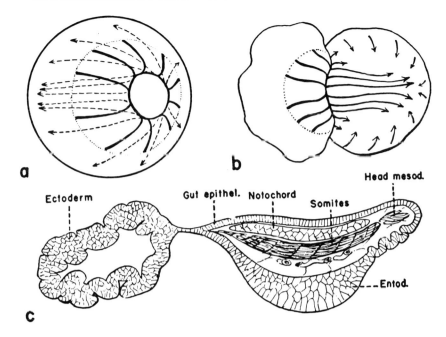

Fig. 6–5. a. Normal gastrula. *Broken arrows* indicate direction of normal invagination. b. Exogastrulation. *Solid arrows* indicate direction of evagination movements. c. Completed exogastrulation. Left, ectoderm. Right, endomesoderm. From J. Holtfreter and Hamburger, 1955.

iments. The eggs of the axolotl are particularly soft; they collapse when the vitelline membrane is removed, and they were the material of choice for Holtfreter's experiment. Total exogastrulation occurred in 17% of the axolotl cases ($n = 167$); the others produced partial exogastrulation. Holtfreter's paper (1933f) deals exclusively with total exogastrulation; partial exogastrulation is discussed briefly in a review article (J. Holtfreter, 1933c).

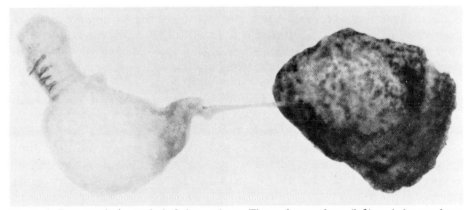

Fig. 6–6. Exogastrula in axolotl, 8-day embryo. The endomesoderm (left) and the ectoderm (right) are connected by a narrow stalk. Note head and gill slits in endomesoderm. From J. Holtfreter, 1933c.

Holtfreter was intrigued by the bizarre features of the "exo-embryo" attached to an empty hull. He found at closer inspection that its different layers were turned inside-out; the endoderm was outside and the mesoderm inside. A great attraction was that he could watch the entire process of exogastrulation, step by step, on the living embryos over a period of several days. Blue and red vital staining marks aided him in the tracing of the complex movements. His fascination with this phenomenon is reflected in the great care and attention to detail devoted to its description and analysis in one of his most elaborate publications (J. Holtfreter, 1933f). Many of the illustrations are by his own hand. I can empathize with my friend, since the prolonged observation of living embryos gave me similar pleasures when I made a study of the motility and hatching behavior of chick embryos. But, in retrospect, the analytical insights which exogastrulation provided seem to be less rewarding than one might have expected.

Holtfreter's preoccupation with exogastrulation attests to his strong interest in the dynamic aspects of embryonic development. In this particular case it reflects also the profound impact which the classical studies of amphibian gastrulation by Vogt exerted on contemporary experimental embryology. His publication almost filled the fifth volume of the Festschrift for Spemann which he edited and presented to him personally on the occasion of his sixtieth birthday in June 1929. Holtfreter's interpretation of exogastrulation leans heavily on Vogt's work; this he fully acknowledges.

Vogt had realized already in 1922 that exogastrulation is not merely the separation of the dorsal from the ventral hemisphere by a constriction in the equatorial plane. He as well as Holtfreter observed the formation of a typical dorsal blastopore, indicating the initiation of an invagination process. In total exogastrulation, reversal sets in almost immediately, and the everted material begins to expand and elongate as it would have done if it had invaginated. Likewise, convergence of the mesoderm occurs, as shown in the midline arrows of Fig. 6–5b. The elongation carries the head and trunk away from the site of the initial blastopore; hence the anterior-posterior axis of the everted endomesoderm runs opposite to the direction which it would have taken normally (Fig. 6–5c). At a later stage, typical head sturctures become recognizable, as for instance a series of gill protrusions separated by furrows (Fig. 6–6). Furthermore, the mesoderm differentiates into notochord and musculature, although the segmentation of the latter into somites is incomplete. Holtfreter was also able to recognize distinct regions of the intestinal epithelium, such as the oral region, pharynx, esophagus, stomach, and midgut. He found kidney tubules at the proper position. Thus, the endomesoderm had accomplished a remarkable feat. It had performed the typical highly integrated gastrulation movements, though in reversed direction, and it had undergone the appropriate histological differentiations. According to Holtfreter, "Despite exogastrulation and the complete absence of the ectodermal covering, the endomesoderm is capable of typical embryo formation" (J. Holtfreter, 1933f, p. 686).

Finally, the ectodermal hull deserves some comments. The complete failure of the isolated ectoderm to proceed with any differentiation, which had been demonstrated earlier for small pieces in the *in vitro* experiment, was now corroborated for the ectoderm as a whole. The ectodermal bag of the exogastrula collapsed and its lining tended to fuse. Only its inherent tendency to expand had materialized: the enlargement of the surface produced irregular folds. Structural differentiation was limited to the formation of cilia. In the absence of mesenchyme, the cells failed to form a typical epithelium; they formed an irregular assemblage of cuboidal or cylindrical cells. Holtfreter notes: "Both [prospective epidermis and prospective neural plate] behave completely identically, dynamically as well as morphologically. Together, they differentiate, belatedly, to epidermis which however retains a primitive character to the end. In the absence of any mesodermal substrate, none of our exogastrulae (28 cases) formed a single nerve cell nor any specialized epidermal structures, such as typical skin epithelium, secretory glands, lens, balancer, cartilage, etc. Only ciliated cells and secreting elements occurred" (J. Holtfreter, 1933f, p. 722).

The barren ectoderm of the exogastrula invites once more—and for the last time—a reflection on Spemann's cherished belief in the "double assurance" of neural induction, by the mesodermal substrate and by a stream of determination in the ectoderm spreading forward from the blastopore. Holtfreter leaves no doubt about his stand in this matter.

> Spemann left the possibility open that the inducing effect of the chorda-mesoderm on the ectoderm could occur not only by the contact established by the substrate but also by neighborhood [contact] on the surface. . . . In our exogastrulae, the substrate [Unterlagerung] was absent; the surface connection of the ectoderm with the dorsal blastoporal lip, however, remained intact on a broad front and for a long period, and it was only broken off approximately at the end of the neurula stage. If a 'stream of organization' emanating from the blastoporal lip exists, then it should have induced the neighboring ectoderm, during this period [to form a neural plate]. But since this never happened, the notion of an inducing stimulus progressing from the blastoporal lip peripherally must be considered as invalidated. (J. Holtfreter, 1933f, pp. 785–786)

How did Spemann react to this challenge? Did he finally relent? Not quite! In his book he gives a concise and fair account of Holtfeter's findings. But when it comes to the touchy point of Holtfreter's dissent, as quoted above, he is not willing to give in:

> One can question whether Holtfreter's conclusions which oppose this assumption [of the spreading of an inductive influence on the surface] are completely compelling. I consider his view as very probable but not as definitely proven. It would seem to me entirely conceivable that the prospective neural plate material of the isolated ectodermal part of the exogastrula had already undergone a labile determination from a certain stage on, but that for mechanical reasons it was not able to execute the morphogenetic movements of neurulation. In this case, one ould not exclude that the determination had spread on the surface from the upper blastoporal lip, hence that this manner of induction would be practicable [gangbar] along with induction by the substrate. (Spemann, 1938, pp. 188–189)

History had decided otherwise. No experimental support for Spemann's view has ever come forth; it has long been laid to rest, and Holtfreter has prevailed.

CHAPTER 7

In Vitro Isolation Experiments (1938, 1965)

The outstanding feature of the organizer, that is, its capacity for complex inductions, culminating in the induction of a whole secondary embryo, has captivated the imagination to such an extent that it has overshadowed another aspect: the properties of the transplanted tissue itself. What does the organizer experiment tell us about the potentialities of the organization center apart from its inductive capacities? Are the mesodermal components of the secondary embryo—the notochord and somites that are derived from the transplant—the result of self-differentiation in the strictest sense? That is, is the notochord derived from the prospective notochord area of the gastrula and the somite musculature from the prospective somite area? Or are the precursor regions more plastic and still interchangeable? At the time when the organizer experiment was done, Vogt's fate maps of the gastrula were not yet available and the exact boundaries of the prospective notochord and somite areas were not known. But even so, the transplants in H. Mangold's experiments were fairly large; they probably included parts of both. The only conclusion that could be drawn was that the transplants had undergone self-differentiation in a general way, that is, they had formed mesodermal derivatives. The specific question raised above was not answered until much later. However, Spemann deserves credit for having raised the problem and for suggesting an experiment that might resolve it. In the discussion of the organizer paper, he states:

> The possibility exists that the implanted piece develops by pure self-differentiation to exactly the same parts which it would have formed at the place from which it was taken. . . . Such a complete self-differentiation of the organizer, however, does not take place; otherwise the implant would be too large for the smaller secondary anlage. To the extent that it fitted in harmoniously its material has been disposed of differently than in normal development. . . . It is not

necessary to assume complete self-differentiation. Distinctly specified developmental tendencies and regulation are not necessarily mutually exclusive. To clarify this point, experiments could serve by which the different parts of the organization center would be tested. . . . If, for instance, a lateral part should later be found in the lateral part of the induced embryonic anlage, this would mean that its laterality had been determined already at the moment of transplantation and that it had retained it after implantation, and had influenced its surroundings correspondingly. (Spemann and H. Mangold, 1924, pp. 632–633)

The implication is, of course, that alternatively the transplant would have undergone an internal regulation and formed the bilaterally symmetrical middle part of the secondary embryo.

At Spemann's suggestion, Bautzmann, who worked at that time in Spemann's laboratory, undertook the transplantation of lateral parts of the organization center (H. Bautzmann, 1926). However, he was interested only in their inductive capacity and therefore used the Einsteck-method, which is not suited for the analysis of the differentiation of the implant. The fact that the induced neural plates were bilaterally symmetrical could mean that the inducing tissue had regulated and become bilaterally symmetrical. But more direct evidence was required to prove the point. Many years later, Spemann assigned to B. Mayer (as his Ph.D. thesis) the experiment of transplanting the lateral part of the organization center to the ventral blastoporal region, which was considered to be a relatively neutral site (Mayer, 1935). The experiment was done heteroplastically between *Triturus cristatus* and *T. taeniatus.* The results were of interest in several respects. When an *early* gastrula was used as the donor, it turned out that indeed the lateral region was not yet committed to laterality; rather, it became the median part of the induced axial system and then complemented itself by assimilative induction of neighboring host mesoderm. Hence, regulation had occurred. The lateral half of a *late* gastrula behaved differently: it retained its laterality and differentiated to notochord and one row of lateral somites. The other row was recruited from host material. While the evidence for regulation in the first experiment seemed to be satisfactory, regulation and assimilative induction were intricately interwoven. A deeper and more detailed analysis of the potentialities of the chordamesoderm district required an experimental design by which the participation and interference of the host embryo was excluded. The step from transplantation to explantation was taken by Holtfreter, after he had perfected his culture medium.

Isolation of Gastrula Parts

Of the many projects started by Holtfreter in the early 1930s, one was the rearing of small fragments of the gastrula in complete isolation. This experiment was performed in 1931 and 1932 on a large scale (over 1,500 explants). However, the results were not published until 1938. The delay was probably due to the higher priority which Holtfreter gave to the investigation of inductions by living and dead abnormal inductors. The explants were taken from the urodeles *Ambystoma mexicanum* (axolotl), *Triturus taeniatus,* and *T. cristatus* (J. Holtfreter, 1938a) and from several frog species (J. Holtfreter, 1938b). I shall deal primarily with the urodeles. Each gastrula was dissected into as many as twenty fragments of different sizes. A dissection scheme is illustrated in Fig. 7–1. As many pieces from a single gastrula as possible were reared together in a single dish, for extended periods, usually for two weeks.

The explants could be grouped in three categories according to their differentiation potentials: the three groups corresponded to the three germ layer derivatives. Isolated fragments of *ectoderm* were essentially incapable of any kind of differentiation. This was in accordance with the behavior of the isolated ectoderm of the exogastrula. In particular, Holtfreter stressed its failure to undergo neuralization: "There is no longer any doubt that there is no evidence for the self-differentiation capacity of the prospective neural plate region. Even the assumption of a labile determination of the neural anlage, which some authors still like to hold in reserve, has no valid experimental foundation" (J. Holtfreter, 1938a, p. 592). The inefficacy of the ectoderm, however, had one notable exception: the gastrula ectoderm of frog embryos frequently produced single or multiple suckers. A pair of adhesive glands or suckers is found on the ventral side of the heads of young tadpoles: with their aid they attach themselves to water plants or other solid surfaces before they begin to swim actively. When that time comes, the suckers are resorbed. Sucker formation occurred in explants from all ectodermal regions, including prospective neural ectoderm. The suckers can be recognized easily by their propensity to secrete clouds of a mucous substance. Holtfreter was intrigued by tiny floating explants consisting almost entirely of a single sucker which trailed a much larger stream of mucus behind it.

The causation of the differentiation of suckers in the explants is obscure. What requires explanation is their indiscriminate formation from any region of the explanted ectoderm, whereas in normal development they are restricted to a single pair at a particular location. Some agents operating in the normal head region reinforce the formation of suckers at their appropriate position and repress them in all other parts of the ectoderm. These findings underscore a point which I have made earlier: that in normal development restricting forces are as important as

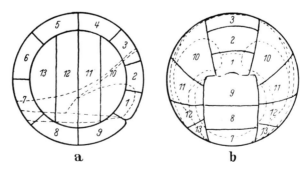

a   b

Fig. 7–1. *In vitro* culture of parts of salamander gastrula. Schema of the isolation procedure. a. Lateral view. b. Posterior view. *Dotted lines* demarcate mesodermal structures; see Fig. 3–7. From J. Holtfreter, 1938a.

activating forces. The integration and patterning of embryonic processes results from the proper balance of the two.

In contrast to the ectoderm, the different districts of the *endoderm* of the gastrula seem to be programmed in considerable detail. Fragments from different endodermal regions differentiated to structures which Holtfreter was able to identify as pharyngeal epithelium, esophagus, stomach, small and large intestine, liver, and pancreas tissue. He states: "Never did I observe a regulative behavior or a deviation from the prospective fate" (J. Holtfreter, 1938a, p. 581), and furthermore: "With respect to the differentiations of the endoderm there exists a great, if not complete, identity with the scheme of the anlagen [in Vogt's maps] both for the axolotl and for *Triturus*" (J. Holtfreter, 1938a, p. 628). But later investigations have led to a revision of this position. T. Okada (1960) and C. Takata (1960), working with the Japanese salamander, *Triturus pyrrhogaster,* used the sandwich method to test the potentialities of different regions of gastrula endoderm when combined with gastrula mesoderm. They found that the mesodermal implants could channel the differentiation of the endoderm fragments in directions that were different from their normal fate. In other words, endoderm was programmed only in a general way; its regional specificity was still sufficiently labile to be redirected by the mesodermal parts. This case signals another warning: that the isolation method, which seems to be ideally suited for the analysis of self-differentiation, has its pitfalls and that overinterpretations must be avoided.

The isolation of the fragments from the dorsal and dorsolateral *mesoderm* district of the early gastrula, that is, essentially prospective notochord and somitic musculature, gave particularly interesting results. I shall single out one isolate which attained an exceptionally high degree of organization. An explant derived from a lateral, prospective somite area had spread on the glass dish rather than floating freely. Although it had originally occupied a lateral position, it regulated and achieved polarity and bilateral symmetry (Fig. 7–2). The notochord rested on a cushion of well-differentiated musculature derived from somitic mesoderm. Muscle cells had migrated out laterally. The mesodermal axial system was complemented by the formation of a central nervous system dorsal to the notochord.

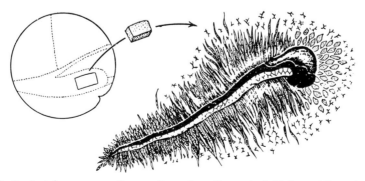

Fig. 7–2. Explant from prospective somite region of gastrula (left) forms bilateral-symmetrical axial organs (right). *Dark,* brain-like structure and spinal cord; *light,* notochord. Muscle fibers have migrated laterally on both sides. From J. Holtfreter and Hamburger, 1955.

It consisted of a well-formed spinal cord and an atypical brain vesicle. In addition, a small mass of epidermis was differentiated near the brain. The remarkable performance of this case exceeds even that of the transplanted organizer in the famous case *Um 132* of H. Mangold's experiment. In the latter, the organizer had at its disposal the host ectoderm in which the central nervous system was induced, and which supplied also the epidermal covering of the secondary embryo, whereas in the isolate these structures had to be fashioned from the material of the explant. This case is of signal importance in that it reaffirms the reductionist conception of the organizer. The creation of a complete axial organ system was achieved by the regulative and transforming capacity of the organization center, operating in complete isolation from the integrating forces of the whole embryo. Nobody will contend that a vitalistic force is needed to explain the virtuosity of the explants.

The majority of the isolates derived from the mesodermal fragments in Fig. 7–1 achieved a much lower level of organization. Polarity was recognizable only in a few instances. Notochord, somitic musculature, epidermis, and neural structures were differentiated in random combinations. Since the prospective notochord and prospective somite regions usually gave rise to both structures in the same explant, the notion of the organization center as a chordamesoderm field seemed to be reaffirmed. The frequent additional differentiation of epidermis and neural structures indicates, however, that this designation is an oversimplification and that we are confronted with new substantive and conceptual difficulties. This point will be taken up in Reevaluation of Basic Concepts.

Fractionation of the Organizer

Four decades after the publication of the organizer experiment, a postscript to the story of the organizer was written which shows it in a new light. In its prime, the organizer had created a second embryo with the aid of the host embryo; now it was removed from its normal milieu and subjected to a radical fractionation process in a reductionist tour de force. It was split first in half, then in one-third, and so forth to ¹⁄₂₄ fragments. Spemann might have considered this treatment as not quite in keeping with his image of the organizer; yet, the intriguing results fully justified this unusual analysis. The experiments were executed by Hiroko B. Holtfreter (Mrs. J. Holtfreter); they earned her a Ph.D. degree, but her thesis was never published. Since the results are of great general interest, they deserve to be presented in some detail.[1]

The design of the experiment was derived from the isolation experiment of gastrula parts by J. Holtfreter discussed in the preceding section. The dorsal region that was subjected to the fragmentation is outlined in Fig. 7–3; it is referred to as the "standard piece." Its normal fate is primarily notochord, but it extends laterally to the somite regions. Very early gastrulae were used throughout. At that

[1]I am grateful to Mrs. Holtfreter for giving her permission to publish this survey of her experiments and for making available a copy of her thesis and of the originals for Figs. 7–3 to 7–7. All quotations are from H. B. Holtfreter, Ph.D. thesis, University of Rochester, 1965. A microfilm of the thesis is available from University Microfilm International, 300 N. Zeeh Rd., Ann Arbor, MI 48106.

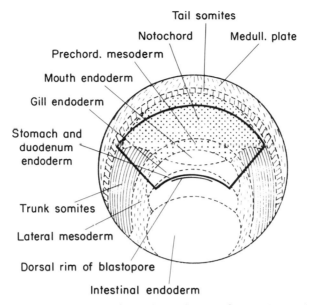

Tail somites

Notochord

Medull. plate

Prechord. mesoderm

Mouth endoderm

Gill endoderm

Stomach and
duodenum
endoderm

Trunk somites

Lateral mesoderm

Dorsal rim of blastopore

Intestinal endoderm

Fig. 7–3. Experiment of fractionation of organizer. Diagram of early salamander gastrula, indicating the dorsal area which was excised, designated as the "standard piece." From H. B. Holtfreter, 1965.

stage, some endoderm, that is, the prospective pharyngeal epithelium, is still located at the outer surface, bordering on the incipient blastopore; this explains the frequent occurrence of pharyngeal epithelium in the explants. The fractionation scheme is presented in Fig. 7–4. The terms "proximal" and "distal" are used to indicate the position of the explants with reference to the blastopore. One notices that in four categories of fragmentations all cuts were made in the proximodistal direction (B, C, D, F). They are referred to as "longitudinal" fragments. In all other categories, "longitudinal" transections were combined with "latitudinal" transections which were perpendicular to the former. When a particular fractionation can be done in two ways (as for example D, E), then the category with longitudinal transections only is given the symbol α and the other the symbol β. The operations were performed with glass needles. With consummate skill, the experimenter managed to rear all or nearly all fragments taken from an individual gastrula in culture medium for 10 to 21 days.

The fragments were reared in three different ways: as so-called *dish cultures,* as *slide cultures,* and in the *coelomic cavity* of partial embryos. Discussion of the latter experiment will be omitted, since it was done only on a small scale. The dish cultures were grown in Holtfreter's solution, to which was added a small amount of Puck's medium (Puck et al., 1958). The explants rounded up and formed irregular masses. All fragments derived from the same gastrula were grown in one dish, in small depressions in the agar bottom in which they floated freely. The slide cultures were actually hanging-drop cultures; they were grown in a small amount of culture medium that was just large enough to cover the explant.

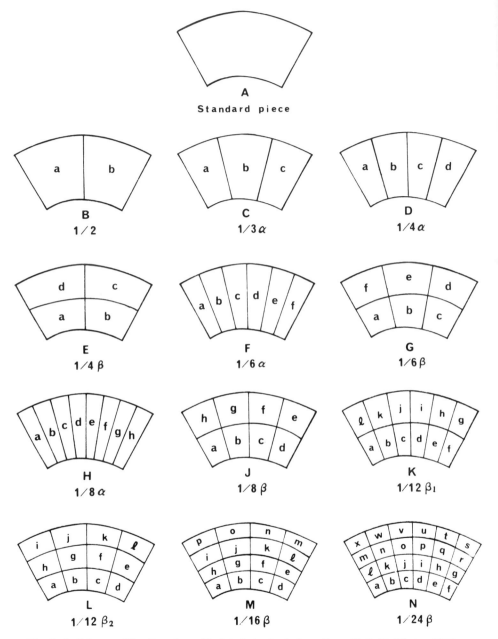

Fig. 7–4. Schema of fractionation of isolated standard piece. From H. B. Holtfreter, 1965.

The coverslip on which the drop was placed was then inverted and mounted on a culture chamber which was kept moist by a film of egg albumen at its bottom. The culture medium was Puck's medium enriched with fetal bovine serum. This method encouraged the migration and spreading of cells on the glass surface, where they formed a monolayer (Fig. 7–5). Antibiotics were added to all cultures,

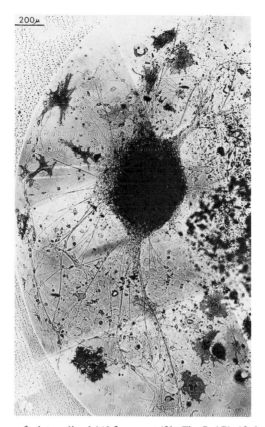

Fig. 7–5. Slide culture of a laterodistal 1/6 fragment (f in Fig. 7–4G), 12 days after explantation. A large mass of neural tissue in the center, surrounded by a flat layer of cells, including nerve cells, muscle cells, and large pigment cells. Note single nerve fibers and fiber bundles radiating from the central mass. From H. B. Holtfreter, 1965.

whereby the mortality was reduced to a minimum. For the dish cultures, the eggs of the Japanese newt, *Triturus pyrrhogaster,* were used. The eggs of this species and those of the American salamander, *Ambystoma tigrinum,* which are much larger than those of the newt, served as the material for the slide cultures.

The experiments were intended originally to answer two basic questions: Is there a correlation between the size of the explants and (a) the diversity and (b) the frequency of the structures derived from them? Not surprisingly, in a general way, the answer was in the affirmative to both questions. But the particulars were very complex. Not only were there exceptions to the rule, but several unexpected findings raised new issues which eventually overshadowed the original questions.

Figure 7–6 presents data concerning both diversity and frequency of the differentiation of a number of structures produced in the dish cultures by the standard piece and by four categories of explants from fractionation. (The corresponding data for the slide culture are very similar.) According to the fate map, the fragments would be expected to give rise primarily to notochord and somitic musculature and to some extent to pharyngeal epithelium. This was, indeed, the case.

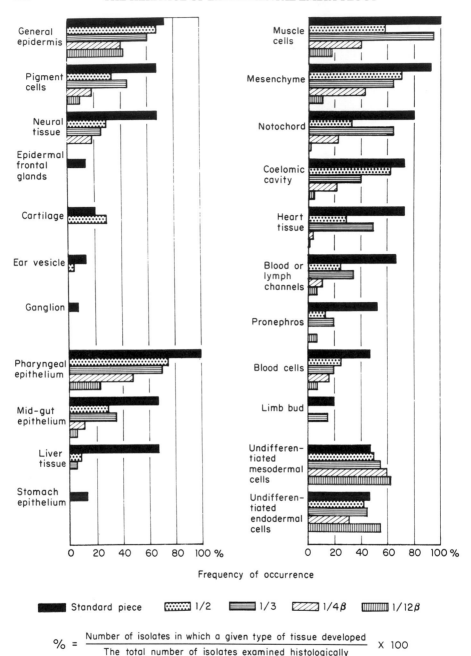

Fig. 7–6. Frequency (in percent) of the differentiations which appeared in dish cultures of the standard piece and four groups of fragments of diminishing size *(Triturus pyrrhogaster).* From H. B. Holtfreter, 1965.

However, surprisingly, there appeared a large number of additional structures that were in no way related to the normal derivatives of the standard piece; they included pigment cells, liver, kidney, heart tissue, and limb buds. Even though the frequency of the unexpected structures was relatively low, their very presence was astonishing indeed. The author's comments are:

> There appeared nearly the whole gamut of mesodermal and endodermal differentiations which are present in the entire larval organism. In addition, practically all the different ectodermal differentiations of normal development could be found. Since according to the operation scheme the ectodermal region was definitely not included in the isolates, the latter differentiations must have emerged from prospective mesodermal (and endodermal) primordia. It is evident then that in response to isolation the dorsal area has manifested the capability of regulative conversion into ectoderm and of transforming the ectoderm into a diversity of structures which normally arise by means of induction (p. 32).

I have some reservations concerning the use of the term "regulative" in this context, and I shall refer to the transformation of prospective mesoderm to epidermis and neural tissue simply as "conversion." At the end of the table are two columns, which designate undifferentiated mesodermal and endodermal cells, respectively. The latter cells can be distinguished by the inclusion of yolk platelets. These undifferentiated cells are healthy and merely arrested in their differentiation. They may account, in part, for the low frequency or absence of certain expected differentiations.

The *decline of frequency* (in percentage) of specific structures with diminishing size of the explants is evident in Fig. 7–6. "The frequency of every one of the listed definable differentiations decreases in direct proportion to the degree of fractionation. . . . The reduction of frequency proceeds along a steeply declining curve, as can be visualized readily if the tips of the five bars pertaining to a particular differentiation are interconnected" (p. 71). The *decline of diversity* is attested by the complete disappearance of certain structures in the smaller fragments, such as cartilage, ear vesicles, and liver.

The stepwise fractionation procedure and the carefully chosen pattern of fragmentation made it possible to extend the analysis an important step further. In a typical morphogenetic field, the potentialities for specific differentiations are expected to be distributed uniformly, and any given region should give rise to the same assortment of structures as any other region. Is the traditional designation of the organizer as a chordamesoderm field, which was based on the results of transplantation of relatively large parts of the organizer, really justified? The fractionation experiments provide an ideal opportunity to test this supposition and to compare fragments of the same size, taken from different locations. This was done on a large scale. The outcome was again more complex than anticipated. While the longitudinal fragments behaved as expected, showing both pluripotentiality and equipotentiality, the differentiations of proximal fragments differed markedly from those of distal fragments. To appreciate the far-reaching implications of the data, I must go into some detail.

The pluripotency of longitudinal fragments was demonstrated in two ways: by comparing the performance of individual segments taken from the same gastrula (Fig. 7–4C,D), or by comparing the sum total of all left and all right fragments in categories subjected to a median transection. H. B. Holtfreter states with reference to the slide cultures of *Triturus pyrrhogaster:*

> Evaluation of the data on the ½, ⅓, and ⅙ fractionation groups led to the same conclusion. Within the set of fractions of each group the wide range of ectodermal, mesodermal and endodermal differentiations occurred at about equal frequencies. Thus these results provided clear evidence that the excised dorsal area had the properties of a morphogenetic field. Its longitudinal subdivisions, ranging from ½ to ⅙ fractions, were each capable of developing into the same broad spectrum of differentiations, a capability which involved regulative mechanisms (p. 143).

The same holds also for longitudinal fragments of *Ambystoma* (p. 151). The rule of longitudinal pluripotency applies not only to the major tissue types but also to structures that occurred infrequently. A particularly impressive case is the presence of rhythmically contracting heart tissue in all three fragments of a fractionation according to Fig. 7–4C.

Another way of testing for pluripotentiality is to compare the performance of all left fragments with that of all right fragments. The result is stated succinctly: "Practically all the cells and tissue types listed [in Fig. 7–6] occurred at strikingly equal frequencies in the right and left sets of fractions irrespective of the type of fractionation" (p. 157).

This is in sharp contrast to the performance of proximal and distal fragments. To simplify matters, I shall focus on the major tissue types: notochord, somitic musculature, pharyngeal epithelium, and epidermis. Proximodistal disparity was not noticed in the ¼ quadrants, but it became increasingly conspicuous in the smaller fragments. A detailed analysis is presented in Fig. 7–7 for fragments of *Ambystoma tigrinum*. Each symbol stands for a 1% representation of a specific structure (total $n = 206$). One notices that notochord, musculature, and pharyngeal epithelium become gradually concentrated in the proximal fragments, whereas epidermis and neural tissue become more frequent in the distal fragments. The following figures give an idea of the overall distribution pattern. If one calculates the percentage representation of specific tissues separately for all proximal and all distal fragments in the six fractionation categories of Fig. 7–7, one finds that notochord, musculature, and pharyngeal epithelium occur at a frequency of 50 to 67% in the proximal and only 13 to 35% in the distal fragments. In contrast, epidermis is represented in 71% of the distal but in only 16% of the proximal fragments. These figures do not reveal one of the most interesting features: that in the smallest fragments of the $\frac{1}{16}$ and $\frac{1}{24}$ fractionations, notochord and muscle differentiations are confined almost entirely to row P_2 and pharyngeal epithelium to row P_1, whereas epidermis and neural tissues are found almost exclusively in row D_2 in Fig. 7–7. In other words, in the smallest fragments the mesodermal structures become segregated almost completely from epidermis and neural tissue.

Obviously these findings are incompatible with the notion of the organizer as a chordamesoderm field. How can one comprehend the paradoxical behavior of the

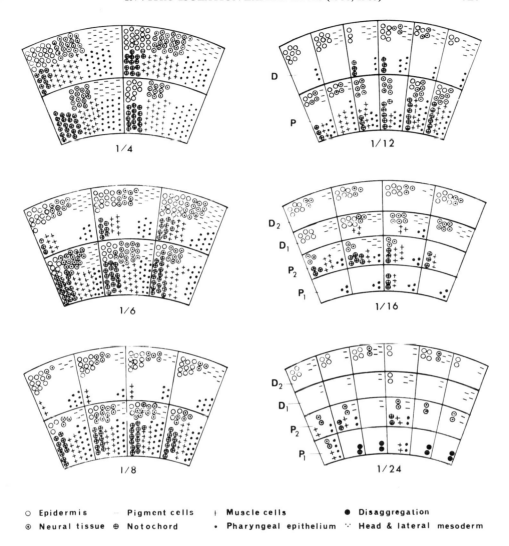

| ○ Epidermis | — Pigment cells | ⊢ Muscle cells | ● Disaggregation |
| ⊚ Neural tissue | ⊕ Notochord | · Pharyngeal epithelium | ∵ Head & lateral mesoderm |

Fig. 7-7. Effects of increasing degree of fractionation of the standard piece on the frequency and regional distribution of six different tissue types in slide cultures of *Ambystoma tigrinum*. A single symbol signifies that the tissue appeared in 1% of the total number of explants of that group. From H. B. Holtfreter, 1965.

distal fragments, that is, the total conversion of organizer material into structures that are normally of ectodermal origin? The explanation offered by the author is plausible in the light of her own and of earlier findings. She reminds us that while the fate map (Fig. 7-3) shows a clear demarcation line between the organizer (mesoderm) and the adjacent prospective neural plate (ectoderm), the boundary is actually still in a state of labile determination in the early gastrula. "With regard to the potencies [of early gastrula regions] there are definitely no distinct borderlines between the prospective ectodermal and mesodermal districts" (p. 192). In fact, defect experiments discussed earlier had shown that *distal* prospective mesoderm can be transformed to neural tissue. On the other hand, Spemann's earliest transplantation experiments on gastrulae (Spemann, 1918) had demonstrated the strong commitment of the *proximal* mesoderm near the blastopore to notochord and somites. The present findings that small fragments taken from the proximal mesoderm region (P_1 and P_2 in Fig. 7–7) produce these structures regularly confirm those of Spemann. Therefore, the author designates this region as "the very core and center of Spemann's organizer" (p. 212). She then argues that "one may conjecture that it is from this core that the properties of the organizer are transferred upon the adjacent cell sheet encompassing in the course of time the whole area of notochord and somites" (p. 209) and furthermore, "It may be assumed that at the early gastrula stage the wave of fate-determination has not yet reached the more distal parts" (p. 212). In other words, in the very early gastrula the cells distal to the organizer core are uncommitted. They become mesodermized by an agent spreading from the core region.

This hypothesis gives a satisfactory answer to the question of why the distal fragments failed to differentiate mesodermal structures. It explains also the field properties of longitudinal fragments, since each of them has a share in the proximal core material. However, it deals only with the special case of the complete ectodermization of small distal fragments, and it leaves two problems unresolved. First, why does the partial conversion of mesoderm to epidermis and neural tissue occur also in the larger fragments, which include proximal organizer core material? And second, why do the small distal fragments frequently differentiate to neural tissue in the absence of a mesodermal inductor? Why do they not form unspecified epidermis in the cases where the mesodermizing agent does not reach them? Moreover, a third point remains problematical: conversion of mesoderm to epidermis and neural tissue does not occur in the organizer experiment. Why does the same region behave differently in the transplantation and the explantation experiment? This key question has profound conceptual implications, which will be dealt with later. At this point we are concerned with the substantive issue of explaining a seemingly paradoxical situation. The main difference between transplantation and explantation lies in the fact that the transplanted chordamesoderm invaginates and continues its differentiation in the internal milieu of the gastrula where it is protected from the external milieu; on the other hand, the explanted mesoderm is exposed to it. This leads to the conjecture that the culture medium, far from being neutral, is implicated in the conversion. In other words, the medium seems to contain an agent that somehow modifies cellular differentiation. In fact, the hypothetical milieu factor(s) seem to play a dual role: it would be responsible for the segregation of the *larger* explants into a mesodermal and an epidermal component. The mesodermal residue in these explants would then operate as an inductor of neural tissue. On the other hand, the neuralization in *small* distal fragments, in the absence of mesoderm, would be ascribed to a direct effect of the milieu on the explant. This is also the opinion of the author. She states: "Possibly some unknown factor contained in the culture medium elicited the formation of these otherwise inductor-dependent differentiations" (p. 214).

This notion is not without a precedent. When fragments of the ectoderm of gastrulae of some amphibian species were grown *in vitro,* they frequently became neuralized. This holds for *Ambystoma maculatum (punctatum)* (Barth, 1941; J. Holtfreter, 1944) and also for *A. tigrinum,* one of the species used in fractionation experiments, but not for different species of *Triturus* (J. Holtfreter, 1944; Yamada, 1950). However, neuralization can be induced in the latter by calcium-free media or by changing their pH; hence there is no fundamental difference between *Ambystoma* and Triturus embryos (J. Holtfreter, 1945). J. Holtfreter has suggested that the "autoneuralization" is caused by a "subcytolytic effect" of the culture medium. By this he meant a mild, reversible cytolysis of some cells of the explant, whereby a neuralizing agent is released to which the intact cells respond. This hypothesis has been challenged, however, and despite many efforts, the role of the tissue culture medium in influencing the differentiation of embryonic cells has not been clarified.

Hiroko Ban Holtfreter was born in Seoul, Korea, which then belonged to Japan. In 1944 she received her B.S. degree from Tokyo Higher Normal School for Women, which after the war became Ochanomizu University. She was then admitted to graduate work in the biology department of Tohoku University in Sendai, Japan. This was a special privilege; generally, women had no access to graduate schools at that time. Through her teacher she became interested in experimental embryology and especially in some problems of differentiation of insect embryos. She was granted a master's degree in 1947 and continued her research in Sendai for another two years, while she held a fellowship of the Japanese Ministry of Education. From 1949 to 1956 she taught at the university from which she had graduated. Her main interest was in research but the postwar conditions in Japan made it very difficult for her to pursue her experimental work. She decided to apply for a fellowship to work in the United States, and in 1956 she received a grant from the American Association of University Women for that purpose.

She had become interested in developing a culture medium for embryonic insect tissues; therefore her natural choice was the laboratory of J. Holtfreter, who had developed a culture medium for embryonic amphibian tissue. She was his research associate from 1957 to 1979 and is now connected with the Department of Microbiology and Immunology of the University of Rochester School of Medicine. She and J. Holtfreter were married in 1959.

Reevaluation of Basic Concepts

The results of the fragmentation experiment necessitate a revision of certain basic concepts in experimental embryology: self-differentiation, regulation, morphogenetic fields (the term that replaced Driesch's harmonious-equipotential system), and embryonic induction. The latter is not pertinent to the explantation experiment, but it will be a major theme in the following chapter. The term "determination," which was widely used by experimental embryologists, is fraught with ambiguities, not the least being its double meaning as a process and a state.

"Determined" is synonymous with "self-differentiating according to the normal fate."

Self-Differentiation

The term "self-differentiation" was coined by Roux in the 1880s. In his ponderous way he defined it as follows:

> The word 'self-differentiation' and its opposite, 'dependent differentiation,' refer to the seat of the causes of a change of an actual or imagined spatially circumscribed formation. If the causes of these changes are located within the thus delimited formation, then I designate the change as self-differentiation of this formation, even though this change requires the input of energy from outside in the form of heat, oxygen, liquid, or solid nutriment. The expression 'self-determination' is supposed to refer only to the 'specific' causes of the change, to the causes of kind and locality and also to the time and intensity of the change. (Roux, 1897, p. 882)

To avoid ambiguities, two parameters have to be identified precisely: (a) the stage of development of the embryo, which defines implicitly the state of development of the embryonic part under consideration, and (b) the precise boundary of that part. It also has to be taken into account that there exist transient states during which a structure is self-differentiating as a whole, while its parts are still plastic and interchangeable. Such systems are referred to as morphogenetic fields. Spemann and many others used the term "self-differentiation" in a more restricted sense than Roux, defining it as differentiation of a given embryonic part according to its normal fate. In modern terminology this would be equivalent to irreversible commitment.

Since embryonic cells usually become self-differentiating before they are visibly (or microscopically) different from other cells, precise criteria have to be established. Spemann specified them as follows: "That the potency for further development is located in the embryonic part itself—which is what is meant by the term self-differentiation—can be ascertained only if it continues to develop according to its prospective fate after isolation. The means [Mittel] of such an isolation are explantation and transplantation" (Spemann and Geinitz, 1927, p. 167). When Spemann started the analysis of the state of determination of gastrula parts in 1915, he opted for the transplantation method. Intuitively or by design, he preferred it to the more radical test by explantation because it would provide the test object with a more normal milieu. For several years, he, O. Mangold, and Geinitz tried the tissue culture method, but without success (Spemann, 1931a, p. 493). In retrospect, their failure was a blessing in disguise. The concept of "self-differentiation" served Spemann as a guide that led him to the discovery of the organization center and to the organizer. His judicious employment of the term spared him the pitfalls that soon came to light.

When the first successful explantations of gastrula parts to the eye chamber and the coelomic cavity were achieved by Kusche, Bautzmann, and Holtfreter (see above, Chapter 6, *In Vivo* Isolation Experiments and a New Culture Medium) and the preliminary results of Holtfreter's *in vitro* experiments became known, it was immediately clear that the two tests, explantation and transplantation, gave conflicting results. Harrison was among the first to grasp the profound theoretical implications of the situation. He addressed them in his contribution to a symposium on "Embryonic Determination" held at the meeting of the American Society of Zoologists in 1932. Referring to the aforementioned isolation experiments and to his own experiments on lens and octocyst differentiation, he stated: "A number of tests involving different conditions may be applied, but they frequently do not give the same answer to the question" (Harrison, 1933, p. 308). Going a step further, he challenged the validity of the concept of self-differentiation: "There is no way of finding out with certainty whether the particular quality which a cell seems to have is finally fixed, for there always may be new conditions, not yet tested, under which other potentialities might be revealed" (Harrison, 1933, p. 318). It follows that if there is no rigorous test for self-differentiation, the concept becomes meaningless. No wonder that the originator of the tissue culture method never used it for the analysis of the differentiation capacities of embryonic structures.

But Harrison's misgivings go beyond methodological concerns. More important theoretical issues were at stake. They are stated succinctly toward the end of his address:

> In dealing with such a complex system as the developing embryo, it is futile to imagine whether a certain organ rudiment is 'determined' and whether some particular feature of its surroundings, to the exclusion of others, 'determines' it. A score of different factors may be involved and their effects most intricately interwoven. In order to resolve this tangle we have to inquire into the manner in which the system under consideration reacts with other parts of the embryo at successive stages of development and under as great a variety of experimental conditions as is possible to impose. Success will be measured by the simplicity, precision and completeness of our descriptions rather than by a specious facility in ascribing causes to particular events. (Harrison, 1933, p. 319)

Harrison's precept had certainly been violated by the earlier search for "the" inductor of the lens or the otocyst or by attributing to the organizer the role of a master regulator. But in the 1930s such oversimplifications were already on the wane. I think what Harrison had in mind in the last sentence of the quotation was the emphasis given by Spemann and others to the *normal* outcome of differentiation as the point of reference, epitomized in the term "self-differentiation." Harrison considered it as a futile enterprise to search for specific chains of causal events that lead to specific end products. He rejected terms like labile determination or double assurance. He perceived differentiation as a process in which a constellation of multiple factors brings about one small step after another. At the same time, the plasticity of the embryonic structures is so great that the substitution of any factor by the experimentalist can derail the system and lead to a different end result that could not have been predicted. Harrison's prophetic warning that the more the experimental designs are varied, the more readily one can expect conflicting results, has never been borne out more dramatically than in the organizer fractionation experiments of H. B. Holtfreter.

I believe that the difference of outlook between Harrison and Spemann was more a matter of emphasis than of fundamental theoretical disagreement, like that between Vogt and Spemann. I see it more as a difference in personal style and temperament. Spemann was more ideological; he envisaged the grand design of vertebrate organization and thought he had found a way of understanding its emergence in terms of a well-orchestrated sequence of inductions and similar actions. Harrison's pragmatic mind was more impressed by the prodigious variety of interactions between parts that was revealed by the experiments and more skeptical of our capacity to decipher their full meaning. At any rate, their differences in perspective did not affect the friendship between them and the great respect they had for each other, which was obvious to those of us who saw them together; nor did they diminish the strength of the influence that both exerted on their contemporaries and followers among embryologists.

The history of the concept which I have scrutinized is not without a touch of irony. In the same year, 1918, Spemann and Harrison published papers that became classics. In the title of Harrison's publication appears the word "self-differentiating" and in the title of Spemann's publication the closely related word "determination." Fifteen years later, Harrison wrote the obituary to both terms. But perhaps it is not uncommon that a concept has its legitimate place in an early phase but becomes obsolete very soon if knowledge grows very rapidly.

Looking back from the vantage point of the isolation and fractionation experiments of the Holtfreters, one can consider them as a vindication of Harrison's standpoint. Yet in no way do they invalidate the conclusions drawn from the organizer experiment. Harrison's broad conception of differentiation can accommodate the results of both experiments. Looking forward, the isolation experiments form the link to more recent experiments which likewise challenge our old views of differentiation; the discovery of transdetermination in imaginal disks of insects (Hadorn, 1965) and the transdifferentiation of embryonic retina cells of the chick to pigment cells and crystalline lens cells (Okada, 1980). And there remains always the much-contested problem of the redirection of cell fate in regeneration.

Regulation

The term "regulation," which designated a basic phenomenon in general physiology, was adopted by experimental embryologists toward the end of the nineteenth century to signify a remarkable feat of embryos: their capacity to restore their organizational unity or the unity of embryonic regions after experimental perturbation. The analysis of regulation began with the classical experiment of Driesch in 1891, in which the first two blastomeres of sea urchin eggs were separated and it was found that each blastomere formed a whole, though smaller, pluteus larva. We remember that Spemann started his scientific career by repeating this experiment in salamander eggs with the same result, which proved that vertebrate embryos possess the same regulative capacity. Regulation can also occur when two sea urchin or salamander eggs in the 2-cell stage are fused; they can give rise to a single oversized embryo. Regulation in later stages of development is illustrated by the replacement of parts of the neural plate after their extirpation and by the reorganization of a lateral part of the organizer to form a bilaterally symmetrical mesodermal axial system (notochord and somites) after transplantation to a ventral position in the gastrula (Mayer, 1935) or after explantation (J. Holtfreter, 1936, 1938a). It is evident from these examples that the concept of regulation is predicated on the more general concept of wholeness, which, as we have seen, is an axiomatic underpinning of embryology.

The traditional view of regulation was challenged by the outcome of the explantation experiments of the organizer. There are several aspects to this issue. I single out first the extraordinary case in which a lateral part of the organizer, when reared *in vitro,* reconstituted a nearly complete axial organ system, including brain and spinal cord (Fig. 7–2). It is true that this case was exceptional and that regulation in the other large explants was less complete. But, fortunately, in experimental embryology a single unequivocal case can prove a point. It is the conversion of part of the mesodermal explants to epidermis and neural tube that creates a conceptual problem. We are no longer dealing with the typical form of regulation, in which wholeness is restored in a self-contained system, but with a new kind of regulation, in which wholeness of the multilayered axial system is restored by one of its components. The mesodermal explant sets aside part of its material to produce a structure, the neural tube, which is not part of its repertory but which is created in the normal embryo and in the organizer transplantation experiment by induction from host ectoderm. To conceptualize this remarkable phenomenon, one could broaden the definition of regulation or, perhaps better, distinguish between two types of regulation: one could refer to the common type as "self-regulation" and to the exceptional cases as "compensatory regulation." The latter term could be applied to neural tube formation by conversion as well as by induction. These terms are meant to articulate the conceptual predicament created by the isolation experiment. I am quite aware that there is no point in adding new words to a dead language.

The incomplete regulation of all other explants can be attributed to several factors. There are, first, the constraints imposed by the culture methods. In the dish cultures, the explants form spherical masses which hardly permit stretching along the main axis. "In the freely floating explants of [J.] Holtfreter and our own, improper execution of the morphogenetic movements results in failure of the tissues to adopt a typical axial pattern" (H. B. Holtfreter, 1965, p. 218). The slide cultures provide a better substrate for axial elongation, but the massive emigration of cells interfered with normal tissue organization. "Being released from the morphogenetically controlling influences of neighboring tissues, the free cells could not establish the histological and anatomical configurations which they would have adopted in normal development" (H. B. Holtfreter, 1965, p. 140). Since axial patterning is thus ruled out as a criterion of regulation, the author suggests two substitutes: "range of diversity of the tissues produced by the explants and degree of anatomical and histological organization of cells and tissues" (H. B. Holtfreter, 1965, p. 216). Many explants of the standard piece and of the longitudinal fragments meet these criteria. But with decreasing size of the isolates, regulation becomes increasingly incomplete and fails altogether when the volume of the fragments falls below a critical mass.

I come to the conclusion that the explantation experiments have revealed a special kind of regulation, but they do not challenge the traditional use of the term.

Morphogenetic fields

Morphogenetic fields are defined as transient embryonic units that are self-differentiating as a whole but regulative within their boundary. This implies that any part can substitute for other parts and that a fragment of adequate size can reconstitute the whole. The forelimb field of a salamander embryo in the early tail bud stage can serve to illustrate other characteristics of fields. They usually extend beyond the topographic limits of the region that actually gives rise to the structure in question. The limb field is larger than the limb-forming cell mass. If the cell material that is destined to build the limb is ablated, adjacent cells move in to close the wound and proceed to replace the limb (Harrison, 1918). Since fields are more extensive than the organ-forming regions, they can overlap. Thus the forelimb field of the early tail bud stage overlaps with the gill and the pronephros (kidney) field. When the limb field district is transplanted to the flank, duplications occur rather frequently, testifying to its regulative capacity. The same point is illustrated by heart duplications, which can be produced experimentally. The heart is formed by two lateral mesodermal cell groups which converge at the ventral midline. If their fusion is prevented, for instance by a midcentral incision, then each cell group forms a complete heart (Ekman, 1925; Copenhaver, 1926). It is generally assumed that the potential for organ formation is highest at the center of the field and that it declines toward the periphery; hence, the term "gradient field" has been introduced. However, the nature of the field quality that is supposed to show a gradation is not known.

The term "transient" in my definition of fields is meant to emphasize that they represent intermediary stages in the continuum of progressive differentiation. They undergo a process referred to as *"self-organization"* or *"segregation."* Subunits arise by cellular interactions; they in turn can display field properties. This process continues until finally every cell group becomes committed to its fate. Such a process has been discussed above in connection with the differentiation of the forebrain-eye field.

The field concept was introduced by A. Gurwitsch (1922) in a theoretical study and by P. Weiss (1925) as a means to interpret his experiments on limb regeneration in adult salamanders. The field concept figured prominently in the 1930s; this is reflected in the textbooks of Weiss (1939) and Huxley and de Beer (1934). The latter authors gave special emphasis to gradient fields. But neither Spemann nor Harrison made much use of the term. Spemann, in a chapter of his book entitled "Embryonic Fields" (1938, p. 297), devoted only a few lines to the limb field. The rest of the chapter is taken up with limb and lens regeneration fields and with the above-mentioned induction fields discovered by J. Holtfreter, which have little in common with embryonic fields. Harrison, in his grand overview, the Silliman Lectures (given in 1949, published posthumously in 1969), and in his Harvard Tercentenary Lecture (1945) mentions morphogenetic fields only in passing in the concluding paragraphs. I suppose that the field concept had no appeal to them because it is an abstraction taken over metaphorically from physics, with no explanatory value. Apparently modern developmental biologists share this view; with a few notable exceptions, they deal with field phenomena without using the term. Later on, I shall try to make a case for its revival. But first I have to raise the question of whether the fractionation experiments present a serious challenge to the field concept.

In this context, I should caution against a possible misunderstanding. The experimental fractionation of the organizer has nothing to do with its segregation in normal development. Fractionation is the experimental isolation of arbitrary fragments at one point in time; segregation is a time-bound process taking place during normal development. Fractionation prevents the cellular interactions which are of the essence in the self-organization of fields. A striking example is the fate of the small distal fragments in the fractionation experiment which were cut off from the mesodermizing influence of the core of the organizer.

The term "organizer" has been used synonymously with "chordamesoderm field." This was justified as long as only the results of the transplantation experiment were available. Since the explanted organizer and its fractions produce a variety of structures besides notochord and somites, the old designation seems to be no longer appropriate. How should one resolve this issue? The answer depends on how one evaluates the explantation experiment as a method of analysis. One can consider the isolation of the organizer and its parts from their normal milieu and their exposure to an artificial medium as a radical treatment which leads to a distinctly abnormal response. One can suggest, furthermore, that we are dealing here with an exceptional case: chordamesoderm as well as other parts of the early gastrula are still in a rather unstable condition, in contrast to other field districts, such as the forebrain-eye field or the limb field, which as constituents of more advanced stages are more stabilized and less liable to convert parts of their cell material to structures that are not included in their normal repertory. From this point of view there would be no compelling need to abandon the general field concept. Alternatively, one can take the stand that Harrison took in the related matter of self-differentiation: if two different experimental methods give different results, and if the definition given above is not universally applicable to fields, then it would be preferable to abandon the term and thus avoid ambiguities.

One point in this debate can be subjected to an experimental test: the behavior of fields other than the chordamesoderm field under comparable conditions of *in vitro* culture. Organ culture of embryonic parts has been practiced widely, but only a few of these experiments contribute critical information to the present issue, because most explantations were done for other purposes. I shall marshal evidence only for the forebrain-eye field and the limb field. The *in vitro* culture of the neural plate of salamander embryos was done by O. Mangold and C. von Woellwarth (1950) and by von Woellwarth (1952). The isolates included anterior and middle parts of the prospective brain region from which the mesodermal substrate had been removed. In all cases only those parts that were expected according to the fate map, that is, well organized fore- and midbrains and eyes, did develop. In addition, a number of structures, such as nose, ear vesicle, and balancers, were induced by the brain parts, and lenses by the optic vesicles.

Of the experiments on limb fields, I consider only those which were done on very early stages preceding the appearance of the limb bud, since the latter demonstrably contains only limb-forming cells. Isolated prospective limb districts of very early chick embryos (0–15 somites) were grown in the coelomic cavity of somewhat older embryos (Rudnick, 1945). The mesodermal component of the transplant formed irregular but identifiable wing and leg cartilages and muscle tissue; the ectodermal covering occasionally formed feather germs and the endodermal layer formed fragments of intestinal tube. The same results were obtained when the prospective leg region was grown on the chorioallantoic membrane of 8-day chick embryos (Murray, 1928).

Admittedly, we have only very limited evidence for my contention that the chordamesoderm field is an exception to the rule that field districts are self-differentiating units which produce only structures that are in accordance with their normal fate. And it would be unwise not to heed Harrison's caveat that no final judgment should be made in matters of differentiation potentials until a wide variety of experimental tests has been explored. Nevertheless, I think that one can make a case for the retention, or rather revival, of the traditional field concept. For one, morphogenetic fields are empirical realities; the concept does not lose its meaning even if the definition given above should turn out to be too restrictive. Furthermore, fields, together with inductions, epitomize the epigenetic mode of vertebrate embryogenesis. And there is one final argument in favor of the plea not to let the explantation experiment stand in the way of reviving the notion of fields. The crucial, and still unresolved, problem posed by the fields, that is, the mechanism of cellular interactions by which fields manage their self-organization to smaller subunits and to terminal differentiations, persists, whether we are dealing with normal or atypical segregation. I hope that cellular and molecular developmental biologists perceive it as a worthwhile challenge. I am aware that analytical efforts in this direction are underway and partly successful. The problem deserves continuous and insistent emphasis.

A final assessment of the impact of the isolation and fractionation experiments on our perception of the organizer leads us in several different directions. First, I do not believe that the new results challenge the conclusions drawn from the organizer experiment and the scores of transplantation experiments in its wake. Second, they underscore a point that I have made repeatedly: that the organizer is part of an integrated system and that its impact on neighboring structures is counterbalanced by the constraints imposed by them on its performance. Third, the experiments have two discoveries to their credit. One is the release of a wide spectrum of potentialities, when the organizer was isolated from its normal milieu, including not only epidermis and neural tissue, but also mesodermal derivatives such as hearts, which are not part of its normal repertory. This shows that the chordamesoderm of the early gastrula is not as firmly committed to its fate as was implied in the transplantation experiments of Spemann. The other discovery is closely related to the same theme. In the early gastrula, the chordamesoderm region is not yet as extensive as the fate map indicates, but is restricted to a core at the upper lip of the incipient blastopore from which a mesodermizing agent spreads to more distal cells. Last, but not least, the isolation experiments have compelled us to subject some basic concepts of experimental embryology to close scrutiny and have taught us to use them with great circumspection.

CHAPTER 8

The Fusion of Biochemistry and Entwicklungsmechanik

The Pioneer Years (1933–1934)

The title of this chapter is taken from J. Needham (1968), to whom I shall refer presently. To witness this event, I have to turn the clock back to the critical year, 1932, which marked the announcement of the induction of neural plate by killed embryonic tissues (H. Bautzmann et al., 1932). It will be remembered that Holtfreter, one of Bautzmann's coauthors, reported that not only dead organizers but also ectoderm and endoderm of the amphibian gastrula, which are not inductors when alive, acquire induction capacity after having been killed by heating, drying, or freezing. This left no doubt that embryonic inductions are mediated by chemical agents. No wonder that biochemically oriented embryologists took up the challenge immediately; the search for inductive substances started off with great vigor and great expectations. Since Wehmeier, like her fellow students in Spemann's laboratory, had no training in biochemistry, she joined forces with the chemist G. Fischer. They obtained neural inductions by organizers that had been treated with alcohol and ether and by substances extracted from amphibian eggs with acetone. The killed organizer tissue was then precipitated with alcohol and—like the other killed tissues—implanted in the blastocoele (Einsteck-method). Wehmeier and Fischer also confirmed the heat stability of the neural inductor (Spemann et al., 1933). Independently of Holtfreter, they found that adult tissues (brain and retina of salamanders) treated with acetone can induce neural plates. At first, they implicated glycogen in neural inductions, but this claim was withdrawn later; the inductions were attributed to impurities in the glycogen used (Fischer and Wehmeier, 1933a,b). At the same time and independently, the eminent chemical embryologist J. Needham, of Cambridge, England, his wife, D. Needham, and the embryologist C. H. Waddington reported neural inductions with cell-free ether extract of salamander neurulae; it was coagulated by heat and implanted in the blastocoele. Similar results were obtained with ether extracts from the viscera of adult salamanders (Waddington et al., 1933). Their first experiments had been done when they were visitors in O. Mangold's laboratory at the Kaiser Wilhelm Institute for Biology in Berlin-Dahlem in the summers of 1932 and 1933. In a more detailed report, a year later, they cautiously suggested that the neural inductor might be a sterol, since the unsaponifiable fraction of ether extracts of neurulae and tail bud stages which had been precipitated with digitonin gave positive results (Needham et al., 1934). Waddington obtained the first neural inductions in the chick embryo by placing coagulated primitive streak (the equivalent of the blastoporal lip of the amphibian gastrula) under the blastoderm of the chick embryo; the blastoderms were grown in tissue culture (Waddington, 1933).

In the same issue of *Naturwissenschaften* in which the Cambridge group published its first results, Holtfreter, after giving more details of experiments with killed embryonic tissues, added two new details of considerable importance. First, he had done what no biochemist would have contemplated but what would suggest itself to a born naturalist: he scanned the entire animal kingdom for tissues with inductive power. Samples taken from cestodes, annelids, insects, fishes, amphibians, reptiles, and mammals proved to have the capacity of inducing neural structures. Second, some of these dead tissues produced complex inductions, including the first mesodermal derivatives, such as notochord, somites, and kidney tubules, as well as rather complete tails. Liver and kidney of adult mice and coagulated chick embryo extract were effective mesoderm inductors (J. Holtfreter, 1933d).

These pioneer experiments started the new era of the fusion of experimental embryology and biochemistry. No fewer than eleven communications on this issue appeared in the inaugural year of 1933: three preliminary reports from Freiburg, three from the Cambridge team, Holtfreter's preliminary report mentioned above (1933d), and his first extensive publication on inductions by killed embryonic tissues (1933e). In addition, the Dutch experimental embryologist M. W. Woerdeman published three articles. On the basis of his investigations of the metabolism of gastrula parts, he suggested a link between glycolysis and organizer activity (Woerdeman, 1933a,b). Furthermore, he had obtained neural and mesodermal inductions in amphibians by rat carcinoma and human muscle (Woerdeman, 1933c).

The preliminary reports were followed in 1934 by extensive publications: one by the Cambridge group (Needham et al., 1934), two by Holtfreter, which will be dealt with in detail later (J. Holtfreter, 1934 b,c), and two Ph.D. theses from Spemann's laboratory by Wehmeier (1934) and W. Krämer (1934). The latter is of interest in that it includes material from Spemann's own experiment of crushing the organizer, which he had reported in 1931. (On several occasions, Spemann turned over sectioned embryos from experiments he had done himself to graduate students who complemented the material with experiments of their own, or merely analyzed Spemann's material, for their Ph.D. theses. As in the case of Krämer, Spemann did not coauthor such publications.) Krämer described several fairly complete secondary embryos which Spemann had obtained with crushed organizer, using the Einsteck-method.

The momentum of the first years was not maintained, however. Several reasons account for the change of pace. The early enthusiam was dampened by the realization that the notion of a specific neuralizing agent was unrealistic. There seemed to be no common denominator to the large number of tissues, live or dead, and cell-free extracts, by which neuralization could be obtained. Some investigators suspected that they might be dealing with simple mechanical stimuli; but this notion was ruled out by numerous experiments with implantations of agar, coagulated egg albumen, gelatine, and celloidine, almost all of which gave negative results. We shall see that some of Holtfreter's findings gave promise to break the deadlock, but this was not realized until later. The situation was aggravated by another fundamental difficulty: the only available bioassay for inductions was the ectoderm of the gastrula. Now it had been shown that killed ectoderm releases a neuralizing agent; hence, there was no way of deciding whether an inductive agent had a direct effect or released the agent, which was present in a masked form. Finally, even if a purified neuralizing substance had been successfully isolated from an abnormal inductor, it would have been impossible to tell whether it had any relation to the neural inductor in the normal embryo. In short, there were sufficient reasons for disillusionment.

The Cambridge group was the only one that continued the work for several years. It culminated in Needham's brilliant synthesis in his book *Biochemistry and Morphogenesis,* which appeared in 1942. (It was reprinted in 1968 with a new introduction.) In the same year, 1942, Needham closed his laboratory, much to the regret of embryologists; he went to China and became completely absorbed in the exploration of the history of Chinese science and technology and its bearing on the meeting of the Eastern and Western minds. In the meantime, Waddington had switched to developmental genetics. Fischer in Germany and Woerdeman in Holland turned to other embryological topics. Wehmeier became incapacitated by illness. Holtfreter left Germany in 1939, shortly before the outbreak of the war, as a *persona non grata* of the Nazi regime. He did not resume publication until 1943. O. Mangold, Spemann's successor in Freiburg, and his colleagues were not interested in the biochemical aspects of induction; they continued in Spemann's earlier tradition.

All this changed after the end of the war. Biochemical embryology became infused with new vitality; a younger generation recaptured the optimistic spirit of the pioneers. Laboratories in different parts of the world began vigorous research activities and broke new ground. I mention the most prominent ones: S. Toivonen and L. Saxén in Helsinki, J. Brachet in Brussels, and T. Yamada in Nagoya, Japan. They were joined in the middle 1950s by H. Tiedemann and his coworkers in West Berlin. The accomplishments of the first decade were presented in a comprehensive survey, *Primary Embryonic Induction,* by Saxén and Toivonen (1962), which had a considerable influence on contemporary thought.

From the large volume of investigations that have accumulated in the meantime I shall select only the small fraction which is concerned with *regional inductor-specificity,* the direct legacy of Spemann. I shall trace the roots of this problem to the 1930s. But before I do this I shall turn for the last time to Spemann himself. I shall try to answer the question that I have been asked frequently: was he disillusioned by the turn of events, by the radical shift from the involvement with the living organizer to the preoccupation with dead organizers? He witnessed only the beginning of this movement. As I have mentioned, he died during the war, in September of 1941.

Spemann's Attitude Toward the Biochemical Approach

Spemann was caught in the whirlwind of the new discoveries by the biochemical embryologists while he prepared the Silliman Lectures, which were delivered at Yale University in 1933. He had received the invitation in the fall of 1931; in the spring of that year he had reported in Utrecht the results of his experiments with crushed organizer which had precipitated the flood of experiments with killed and denatured inductors. His intent was to give in the lectures at Yale an overview of his life's work and of that of his school from his organismic point of view. This is clearly stated in the introduction. There he contrasts the reductionist approach, "whereby one allows physical and chemical processes to act on the developing organism," with the causal-analytical experimentation on the living embryo, "whereby one tries to disassemble the total process first into larger partial processes and progressively analyzes them further. In doing so, one remains for the present entirely in the sphere of the vital and advances step by step, unconcerned about the question of how far the resolution into the non-vital will be possible." He agrees that "both ways are practicable and in the final analysis have the same goal" (1938, p. 2). His personal inclination and talents made him choose the second road.

But now the young biochemical embryologists had handed him an abundance of data which he had no intention of disregarding. The lectures, which appeared in book form in 1936 in German and in 1938 in English translation, contain two chapters, on "Induction by abnormal inductors" and on "Means [Mittel] of induction," which discuss in detail all pertinent biochemical experiments up to 1935. How did he fit these data into his overall frame of reference? On one level he shared everybody else's disappointment "that the hope which was well founded, when the inductive capacity of dead implants was discovered,—the hope that the exact analysis of the means of induction in the special case of neural plate induction by dead substances might lead to unequivocal information on the means by which the neural plate is determined in normal development" (1938, p. 236) was not fulfilled. On a deeper level, he had to come to terms with the inner conflict between his organismic and the reductionist view. This was greatly facilitated by one aspect of the experiments with abnormal inductors which the practitioners of the new approach had hardly become aware of: the shift of emphasis from the embryonic inductor to the reacting system, the gastrula ectoderm.

There had always been a delicate balance between the roles assigned to the inductor and to the induced structure in the induction process. The term "organizer" implies a strategic role of the inductor, and the organizer experiment vindicates this view. But Spemann had never lost sight of the contribution of the responding tissue to the end result. His publication of 1921 on heteroplastic transplantation ends as follows: "The results of heteroplastic transplantation make it probable that the determinative influence is preponderantly a releasing one and that the organizing activity, at any rate during early development, is not an "instructing" one but one that disposes [disponierend]. The necessary "instruction" is brought along by the cells as their endowment" (1921, p. 569). The xenoplastic transplantation experiments of Schotté (Spemann and Schotté, 1932) are a striking illustration of this point. Now this issue came into sharper focus than ever before. If substances released from dead tissues can evoke the formation of brains or even axial systems, then the competence of the gastrula ectoderm moves to the center of the stage. The response of the gastrula ectoderm is akin to the self-organization of morphogenetic fields, except that in the case of the gastrula ectoderm it is initiated by an extrinsic stimulus. Pointing out analogies of this kind may be useful one day, even if at the moment this does not help to understand the mechanism involved.

Of course, the discoverer of the organizer did not fail to realize the full force of the conceptual dilemma posed by the abnormal inductors. He stated his position very clearly in the concluding pages of his book:

> The inducing effect of the organizer has been considered by me from the beginning as a releasing one. Furthermore, from the beginning the question was raised: what is the share of the action system and the reaction system in the origin and the character of the induction product? In the experiments undertaken to solve this problem the share of the reacting system turned out to be greater and greater and eventually so great that the organizer concept itself became problematical. If the most diverse tissue types, even of warm-blooded animals can produce inductions in *Triturus* as the host, and if the formation of complex morphological structures can be evoked by simple substances, then the entire complexity must be attributed to the reacting system, and the concept of the organizer is no longer fitting. A 'dead organizer' is a contradiction in itself. [A detractor could easily lift the last two sentences out of context and make it appear that Spemann himself discredited the experiment that had won him the Nobel Prize. However, he continued:] On the other hand, one cannot do away with the experimental result that a [transplanted] piece of the upper blastoporal lip invaginates in a direction appropriate to its intrinsic structure, even though it may be opposite to the axes of the host embryo; that it supplements itself by incorporating adjacent mesoderm [of the host] to form a complete axial system; that a piece of archenteron roof [invaginated organizer] when implanted into the blastocoele and thus placed in contact with ectoderm induces a neural plate which can be oriented in opposite direction to the primary [neural plate] and can add lenses and ear vesicles in appropriate order and proportion. Here the entire [axial] orientation and order of the processes depends on the organizing effect of the transplanted piece of the upper blastoporal lip; its position determines what shall become of every cell of the reacting system exposed to its influence.
> Obviously, the induction by an inductor lacking a morphological structure differs from one caused by a living organizer in one important point: in that the [axial] orientation and character of the processes is accounted for entirely by the reacting system, its structure and its other properties. (1938, pp. 369–370)

Spemann can hardly be blamed for not telling us how he envisaged the origin of patterned differentiation in the gastrula ectoderm in response to abnormal inductors. It would be difficult even today to answer this question. But the quotation indicates that he had no difficulty in reconciling the new data with his basic position and that he did not consider them as a challenge to the organizer concept. He had always realized that embryonic induction poses three problems: the role of the inductor, the role of the reacting system, and the transmission of the inductive stimulus. His own methods permitted him to analyze the first two, but the third had remained elusive. He recognized that the new biochemical approach opened the door to the analysis of the mechanism of transmission. At the beginning of the chapter on "Means of Induction" he states:

> Very recently, the question of the means of induction has been given the most intensive consideration. When the work on this book began four years ago, the problem was posed and the way to its solution was indicated. But none of the experiments then undertaken had given a positive result. Shortly thereafter successes began to appear, however; the physiological chemists took possession of the subject, and now a bewildering abundance of factual data is available. Many of them permit an exact decision between alternatives proposed earlier; others are so surprising and revolutionary that entirely new possibilities have come into view. (1938, p. 222)

This is not the voice of a defeatist.

However, this optimistic outlook cannot conceal the fact that a major issue continued to perplex him. He could accept the chemical nature of the inductive agent, but he could not separate this reality entirely from the effect of the agent on the reacting tissue. I quote from the last paragraph of the chapter on "Means of Induction":

> Every insight won [by the chemical approach] reflects on the critical assessment of the most general questions [of development]. Hence now the important question arises: to what extent can something without structure in the morphological sense, such as a chemical compound, cause the origin of structure? Therefore, it will be one of the next tasks to be undertaken by us to repeat the older experiments on [axial] direction and regional specificity of induction by using artificial inductors. (1938, pp. 245–46)

We remember that he means by "structure" axial polarity, bilateral symmetry, regional specificity and other patterned differentiations. In a previous quotation he had taken the position that in the case of abnormal inductors the responsibility for the organizational pattern in the induced structures rests *exclusively* with the reacting system, that is, the gastrula ectoderm. But now, this issue seems to have been reopened and another possibility given consideration: that abnormal inductors might have some influence on the regional specificity of the end result. At least that is my interpretation of the juxtaposition of the terms "regional specificity" and "artificial inductors." If I am correct, then he has anticipated—however vaguely—one of the most important results of the subsequent analysis of inductive agents: that, indeed, some animal tissues and tissue extracts induce specifically forebrain and other head structures and others induce trunk-tail structures. Spemann himself did not pursue his suggestion actively, but his last discovery, the identification of head organizers and trunk-tail organizers, paved the way for the later discoveries. It is therefore appropriate to devote the last chapter to the general theme of regional induction-specificity of abnormal inductors.

CHAPTER 9

Regional Induction-Specificity of Heterogenous Inductors

The most advanced case of H. Mangold's organizer experiment showed the remarkable capacity of a small organizer transplant to produce a secondary embryo; but it had one blemish: it lacked the anterior part of the head, including forebrain and eyes. It was suspected that the deficiency was due to the fact that the transplant had been taken from an *advanced* gastrula; it had not included the anterior brain-inducing part of the organizer, which had already invaginated and moved to a rostral position. If this interpretation is correct, then one should obtain complete secondary embryos by transplanting the upper blastoporal lip of an *early* gastrula. The experiment was done by Spemann (1931a, p. 397) and others, with the expected result. To test the region-specificity of the organizer in a rigorous analytical experiment, Spemann undertook an extensive series of experiments in which anterior and posterior parts of the organizer were transplanted to different regions of the host. The hypothesis was confirmed, and Spemann introduced the terms "head organizer" and "trunk organizer" (Spemann, 1931a). Trunk-tail organizer would perhaps be a better designation for the latter.

O. Mangold (1933) gave a very impressive demonstration of region-specific inductivity of the organizer, using a different experimental design. In early neurulae of *Triturus* the neural plate was carefully removed and the underlying mesoderm mantle (archenteron roof) exposed. The latter was subdivided into four equal parts along the main axis, and the individual fragments were implanted into the blastocoele of gastrulae (Einsteck-method). Large induced structures protruded from the belly region of the host; most of them were recognizable as head, trunk, or tail differentiations (Fig. 9-1). The first segment, which in the embryo lies in front of the brain, induced only balancers, mesenchyme, and sensory epithelium. The second quarter regularly induced heads with typical brain vesicles, eyes, noses, and ear vesicles. The third quarter induced primarily hindbrain, spinal cord, and musculature, and the fourth segment induced spinal cord, somites, pronephros, and tails. However, there was some overlap; for instance,

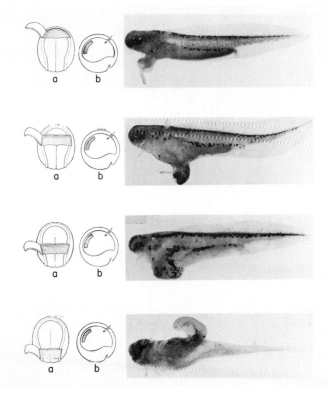

Fig. 9–1. Regional induction-specificity of the four quarters of the mesoderm (archenteron roof) of the salamander neurula, demonstrated by the Einsteck-method. From O. Mangold, 1933.

the third segment occasionally induced eyes and balancers, in addition to its appropriate inductions. This is another instance of overlapping induction fields, which we have encountered earlier in a very different experiment of Holtfreter's.

In the same publication, O. Mangold reported a parallel series of experiments in which he tested the inductivity of the four consecutive segments of the neural plate, using again the Einsteck-method. They displayed nearly the same regional induction specificities as their mesodermal counterparts from which they had received their regional specificity at an earlier phase of development. This experiment is particularly relevant to our main topic, because the neural plate is unquestionably an abnormal inductor. Hence this experiment can be considered to be the first demonstration of region-specificity of abnormal inductors. Mangold did not pursue this topic any further beyond the brief preliminary account.

Neural and Mesodermal Inductions by Abnormal Inductors

Breaking new ground (Holtfreter, 1933–34)

Spemann's quest for "a repetition of the older experiments on regional differences in induction with artificial inductors" (1938, p. 246) became the leitmotiv for an intense research effort, the beginning of which shall be traced in the following pages. As in most other instances in experimental embryology, chance observations by the prepared mind were as effective as carefully designed experimental paradigms.

Almost all inductions obtained by the biochemical embryologists in the early period were neural tissue or brain vesicles, with no clear sign of regional characteristics. It seemed then that the abnormal inductors had a general, nonspecific neuralizing effect. This led the Cambridge group to an interesting theoretical speculation. They postulated a dual nature of determination:

> In . . . determination, two processes, or better, two aspects of the same complex process, can be distinguished. Roughly speaking, these two processes can be called 1) the determination that an embryonic axis shall be developed and 2) the determination of the character of the axis. . . . We suggest that the first type of determination be spoken of as *Evocation,* since it consists in the evoking of an embryonic axis from the competent ectoderm, and that the second be spoken of as *Individuation.* . . . In the ordinary induction, one of these determinations, evocation, is always performed by the graft, while the other, individuation, is performed by the graft and the host working together, either in a cooperative or an antagonistic manner. . . . It is probable that no individuation can be performed by dead organizers or organizer extracts, but for the present the question must remain open. (Needham et al., 1934, pp. 408–409)

The idea of a dual nature of induction was revived later by the Dutch experimental embryologist P.D. Nieuwkoop (Nieuwkoop et al., 1952), who on the basis of his own experiments distinguished between "activation" and "transformation." The notion of a two-step process, whereby the second step specifies the regional characteristics, has not found wide acceptance, and I shall not pursue it any further. One point, namely, the assertion of Needham that individuation requires the participation of a living host embryo, has been refuted by many experiments, which will be dealt with below. Altogether this theoretical construct led to a dead-end road.

The first guidepost that pointed in the right direction was caught sight of by Holtfreter in experiments that had no such theoretical basis and no direct relation to regional inductor-specificity. His way of designing experiments was also different from that of Spemann, who always took one analytical step at a time and addressed specific questions to the embryo, expecting it to give an unequivocal answer. Holtfreter would have in the back of his mind a few general questions which interested him particularly. He would then either do an appropriate experiment or, as in the present case, cast a wide net, confident that the catch would include some novelties valuable to him. I refer to his large-scale experiments of testing the inductivity of gastrula and neurula parts that had been killed in different ways, and to his search for inductive tissues in different vertebrates and invertebrates.

The first clue to an understanding of regional induction-specificity was obtained in the Einsteck-experiments with ecto-, meso-, and endodermal embryonic parts killed by heat treatment at different temperatures, and with alcohol and ether extracts of these parts (J. Holtfreter, 1934b). The very first case described in detail made an important point. The implant, the *posterior region* of a neural plate from an embryo which had been killed by heating at 60°C but not dried, induced a bulky protrusion in the heart region of the host embryo. Two balancers indicated that one was dealing with the induction of head structures. This was confirmed by the microscope sections, which showed induced brain vesicles, a nose, and frontal glands. Holtfreter did not fail to realize the significance of these findings. "It is remarkable that this dead implant, which originated from the caudal neural plate, induced only brain-like and mesectodermal tissues [pigment cells and mesenchyme], whereas the same piece, when alive, used to induce predominantly tail organs (Mangold, 1933). The capacity to induce caudal spinal cord and meso-dermal organs seems to have been lost by heating and transformed to the capacity to induce rostral organs [brain, nose, balancers]" (J. Holtfreter, 1934b, p. 233). I consider this astute observation as a significant conceptual breakthrough. It is the first step toward the identification of two distinct chemical agents mediating the action of Spemann's head- and trunk-tail organizer, respectively, the former being heat-stable and the latter inactivated by heat. Before I elaborate on this point, I mention another important finding: the first cases of induction of *mesodermal* structures, that is, kidney tubules and musculature, by abnormal inductors. So far, a great variety of dead embryonic and adult tissues, cell-free extracts, and chem-ical substances had induced exclusively neural structures, and the opinion was generally held that all abnormal inductors are neuralizing agents. At first it was thought that Holtfreter's finding was just a minor exception to this rule, since he had obtained only six mesodermal inductions in over 600 experimental embryos, that is, less than 1%. Later, it was found that mesodermal inductivity is wide-spread. The mesodermal inductions were derived from killed ectoderm, neural plate, organizer, and endoderm. Hence the mesoderm-inducing agent is not restricted to any special structure, nor is it limited to any special mode of killing. It is not clear why in these cases the agent escaped its normal fate of being inac-tivated by the different killing methods employed.

The idea expressed in the quotation above—that a spinal cord- and mesoderm-inducing agent is transformed into an agent that induces rostral head structures—was corrected in the discussion of the same paper in a section with the intriguing title "Several inducing substances?" After a discussion of all the neural inductions obtained by the Freiburg and Cambridge groups, Holtfreter states:

> However, we have certain indications that also in the *Triturus* embryo there occur different inductive substances. First, there is the regional induction-spec-ificity of the living inductors. This is expressed by the induction of voluminous brain-like neural parts and sense organs by anterior regions of the substrate and of thin neural tubes without sense organs but frequently with mesoderm by pos-terior regions. Second, after a number of killing procedures, the mesodermal inductivity is abolished almost completely, whereas the neural induction capac-ity remains intact. On the other hand, tissue extracts and synthetic substances have induced so far only neural tissue. Hence it appears that at least the meso-dermal induction capacity is based on substances that are different from the neural one. Furthermore, the induction of lenses, balancers, frontal glands, and others, could be due to other specific substances. (J. Holtfreter, 1934b, pp. 296–297)

Obviously, Holtfreter had a clear conception of the existence of specific, chemically different inductive substances, but he failed to take the decisive step: to relate this notion to regional induction-specificity. Although he had overwhelming evidence that the neural inductions that occurred in nearly 100% of his experimental embryos represented brain parts and sense organs, and only rarely "thin neural tubes" that could be considered as spinal cords, he did not draw the conclusion that seems obvious in retrospect: that the heat-stable agent is actually the inductor of the forebrain and the heat-sensitive agent the spinocaudal inductor. He tried to explain the rarity of spinal cord inductions differently: "The almost complete absence of elongated neural tubes is due, in part, to the fact that the dead inductors themselves cannot grow in length, as is the case with living notochord and musculature" (J. Holtfreter, 1934b, p. 287). Later in the discussion he advanced another hypothesis: that the anterior part of the organizer produces a larger quantity of the neuralizing agent than the posterior part. This would result in a broader neural plate in the head region and a narrower one in the trunk-tail region. But these erroneous speculations should not detract from the fact that Holtfreter provided the first experimental proof for the distinction between a heat-stable neuralizing agent and a heat-sensitive mesodermizing agent; this became the foundation for all subsequent work on regional inductor-specificity.

The report on the experiments with killed embryonic tissues was followed in the same issue of *Roux' Archiv* by a comprehensive account of Holtfreter's search for inductive tissues in invertebrates and vertebrates (J. Holtfreter, 1934c). These experiments established the remarkable fact that many adult tissues, alive or killed, contain inductive substances to which gastrula ectoderm is responsive. The great variety of inductions, including nearly all organs and structures that can be found in an embyro, is as impressive as the high degree of organization attained in many instances (Fig. 9–2). The inductions range from nearly complete secondary embryos to isolated lenses and balancers. No correlation was found between the inducing tissue types and the tissues induced by them. I shall concentrate later on those findings which are pertinent to the central theme of regional specificity. But first, tribute should be paid again to Spemann and O. Mangold, the originators of the Einsteck-method, which made this analysis possible. Fortunately, the one drawback of the method—that the experimenter has no control over the exact position of the implant on the ventral side of the host embryo—has no serious consequences; the axial organization of the host embryo has little influence on the outcome. Second, a general point should be made: neural inductions, and, in particular, brain vesicles, appeared in 100% of the experimental embryos, providing further evidence for the notion that a neuralizing agent is very widespread, and for the corollary that the gastrula ectoderm is readily disposed to respond to external stimuli with neuralization. The latter idea fits in with the observation mentioned earlier that the gastrula ectoderm of some amphibian species becomes neuralized simply by exposure to an inorganic culture medium. Of considerable theoretical and practical interest is the finding that some adult tissues, such as mouse liver and kidney, and also chick embryo extract are very effective inductors of mesodermal derivatives such as notochord, musculature, kidney tubules, and tails. Spinal cords occurred only in combination with notochord and somites; they were obviously induced by the mesoderm. Hence, the idea emerged gradually

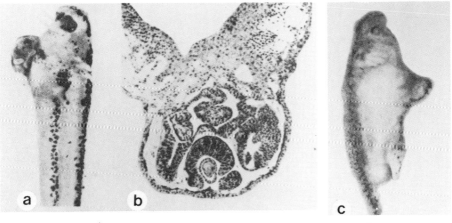

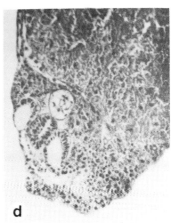

Fig. 9–2. Highly organized inductions by heterogenous inductors. a,b. Induction of anterior head structures (brain, eye with lens) by implantation of heat-killed anterior neural plate into the blastocoele of an early gastrula of *Triturus alpestris*. c. Induction of an incomplete secondary embryo by implantation of boiled human thyroid gland into an early gastrula of *T. alpestris*. d. Cross-section through posterior head region of secondary embryo (c), showing hindbrain with a pair of ear vesicles, notochord, and somites. From J. Holtfreter and Hamburger, 1955, after J. Holtfreter, 1933e, 1934c.

that a neuralizing agent provided for forebrain induction and a mesodermizing agent is indirectly responsible for spinal cord induction. However, this inference was not yet drawn by Holtfreter at that time. (The remaining question of how the hindbrain comes into existence will be answered later.) Finally, a few experiments were done in which fresh as well as boiled mouse kidney was used as inductors. As expected, the former induced mesodermal structures, such as muscle and kidney tubules in combination with brain, but the latter induced brain vesicle only.

This extensive set of experiments seems to contribute little to the problem of organization along the axis. On the contrary, there were a number of cases in which head structures such as brain vesicles and balancers and distinct tails occurred side by side (Fig. 9–3); this seemed to confound any notion of regionality. But here the conception of two inductive substances, one for brain and one for trunk-tail, proved to be helpful: both could be present in the same tissue and released simultaneously.

If these experiments were of no great direct benefit to the analysis of region-specific inductivity, they provided the foundation for all future biochemical work on this and other problems. Holtfreter states in the introduction of this paper:

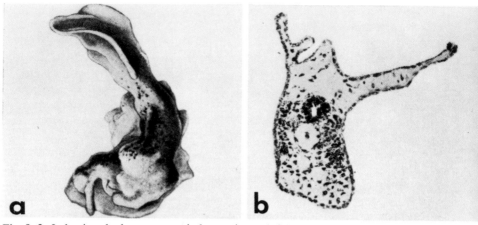

Fig. 9–3. Inductions by heterogenous inductors in sandwich experiment. a. Induction of balancer and tail by fresh salamander liver implanted into vesicle of gastrula ectoderm of *Triturus alpestris*. b. Induction of spinal cord, notochord, somites, and tail fin by fresh salamander liver implanted into a vesicle of gastrula ectoderm of *T. alpestris*. From Chuang, 1938.

"Our experiments had also a particular practical interest. It was the desire to find an inducing tissue which can be procured easily and in ample quantity and therefore is better suited as the starting substance for the chemical-physical analysis of the inductive substances than the tiny normal inductors of the *Triturus* embryo, which are laborious to isolate. The experiments were successful also in this respect" (J. Holtfreter, 1934c, p. 309). Indeed, the fractionation and purification of inductive substances would not have been possible without the availability of large amounts of starting material. However, this gold mine was not exploited until fifteen years later.

How many region-specific inductors? (Chuang and Toivonen, 1938–40)

A few years after the publication of Holtfreter's experiments that indicated the existence of two qualitatively different inductive agents in adult tissues, a neuralizing and a mesodermizing agent, two gifted young investigators took up this topic and carried the analysis a few steps further. One, H. H. Chuang, was a Ph.D. candidate in Holtfreter's laboratory at the Zoology Department of the University of Munich; the other, S. Toivonen, was a doctoral candidate in zoology at the University of Helsinki. How was experimental embryology transplanted to Finland? Toivonen's mentor, the experimental embryologist G. Ekman, had worked in the laboratory of the German anatomist, H. Braus, a close friend of Spemann. After the death of Braus in 1924, Ekman spent some time in Spemann's laboratory in Freiburg, where I made his acquaintance. I have mentioned his experiment on heart duplication. The topic of Toivonen's thesis was suggested by Ekman, who had been intrigued by Holtfreter's observation of the induction of a few isolated lenses in the absence of the optic vesicle. Toivonen was supposed to search for a lens-specific inductor. As luck would have it, he actually found one in the thymus of the guinea pig. But he also obtained a wide variety of complex head and trunk-tail inductions by other adult tissues, and he broadened the scope of his thesis accordingly. In this he was lucky again; Rotmann (1942) showed a few years later that the thymus is by no means a lens-specific inductor; it induces other structures as well. After the untimely death of Ekman in 1937, Toivonen continued on his own. He later became a leading figure in his field. Both Chuang and Toivonen published preliminary reports in 1938. Chuang's major publications appeared in 1939 and 1940, that of Toivonen in 1940. (Fortunately, his 150-page thesis was written in German and not in Finnish.) The fact that the two investigators were not aware of each other's work and obtained similar results contributed to the immediate impact of their discoveries.

Before I deal with their findings, I refer briefly to a matter of terminology. In 1942 Lehmann wrote a thoughtful review on inductor specificity in which he acquainted the experimental embryologists with two useful terms that had been coined much earlier by comparative anatomists and embryologists of the brain. The term *archencephalic* refers to the embryonic forebrain, which gives rise to the telencephalon and diencephalon. The term *deuterencephalic* refers to the embryonic midbrain and hindbrain, which form the mesencephalon, metencephalon, and rhombencephalon (medulla oblongata). Strictly speaking, these terms identify brain parts, but their meaning was broadened by experimental embryologists to include topographically related head structures. That is, archencephalic also stands for eyes, lenses, noses, and balancers, and deuterencephalic also refers to ear vesicles.

One should also distinguish between primary and secondary or direct and indirect inductions by adult tissues. For instance, brains are primary inductions, resulting from the direct contact of the gastrula ectoderm with the abnormal inductor. They, in turn, induce secondarily noses, lenses, ear vesicles, and balancers. But the occurrence of isolated lenses warns us that it is not always possible to distinguish between the two modes of origin. On the basis of the organizer experiment and other experiments, one can assume that spinal cords in trunk-tail inductions are secondarily induced by the chordamesoderm; and the dorsal fins are induced, in turn, by the spinal cord. But again, it is not always possible to ascertain that this is actually the chain of events.

I shall take up first Chuang's three publications (1938, 1939, 1940). He selected two tissues which Holtfreter had found to be effective inductors: salamander liver and mouse kidney. They were implanted fresh or boiled for different lengths of time, and both the sandwich method and the Einsteck-method were used. The experiments were done on a large scale, permitting the statistical evaluation of the data.

It should be understood that Chuang's experiments were intended primarily to analyze the influence of different regions of the host embryo on the type of inductions, in the Einsteck-experiment. The regional specificity of the inductors was only of secondary interest to him. The explantation (sandwich) experiments were meant to be only controls, yet they gave the most interesting results from our viewpoint. The fresh salamander liver induced very complex structures: a very high percentage of brain vesicles, a somewhat lower percentage of eyes, noses, ear vesicles, and balancers, and, in addition, notochord, somitic musculature, kidney tubules, and tails in 40 to 50% of the cases (Fig. 9–3). The brain vesicles were frequently irregular, but the adjacent sense organs (eyes, ear vesicles) identified them as either archencephalic or deuterencephalic or both. In contrast, the mouse kidney did not induce any mesodermal tissues; brain vesicles occurred in 100% of the cases, and sense organs and balancers regularly, though not quite as frequently. Thus Holtfreter's contention that some adult tissues contain only a neuralizing agent and others both a neuralizing and a mesodermizing agent, which had been based on only a few cases, was now confirmed in a very substantial material. Of equal importance is the fact that Holtfreter had used the Einsteck-method. Chuang's isolation experiments provided clear evidence that mesodermal structures can be evoked directly in gastrula ectoderm, independently of a host influence and of the availability of host mesoderm. Another observation points in the same direction. Whereas the brain vesicles were irregular in most instances and difficult to identify, there were a few cases in which the induced structures showed a high degree of organization, including bilateral symmetry, such as, the differentiation of a typical hindbrain, flanked by two ear vesicles, or a tail with a normal set of axial organs and a fin (Fig. 9–3). These cases refute the claim of Needham and Waddington that region-specific patterning of this kind, which they referred to as "individuation," requires the participation of a host embryo. Obviously, the gastrula ectoderm is capable of a high degree of self-organization.

Using the Einsteck-method, Chuang confirmed that the mesodermizing agent is heat-labile, whereas the neuralizing agent is heat stable. Dipping the liver tissue into boiling water for a few seconds was sufficient to inactivate the mesodermizing agent. "I conclude that we are dealing with qualitatively different induction factors, whereby boiling destroys the mesodermal factor more readily than the ectodermal factor" (Chuang, 1940, p. 35).

The perplexing sight of balancers and brain side by side with a tail, in a small explant, as in Fig. 9–3a, should confound any idea of the creation of the pattern of organs along the main axis of a normal embryo by region-specific inductors. The apparent paradox is readily resolved, however, by the simple and now well-documented assumption that the implant contains an archencephalic and a mesodermizing agent, both in an active state.

I have mentioned that Chuang's main concern was the influence of the different regions of the host embryo (the head, heart, anterior, and posterior trunk levels) on the outcome of the inductions. This problem has intrigued experimental embryologists ever since it was discovered by Spemann (1931a). Using the Einsteck-method, he had found that whereas head organizers induce head structures at any level, trunk organizers induce trunk-tail structures at the trunk region but heads at the head level of the host. This seemed to indicate that as a rule the intrinsic region-specificity of the inductor prevailed but that in some instances forces emanating from the host embryo play a decisive role. But how can the role of the host embryo be explained? Spemann gave much thought to this intriguing question. In the lengthy discussion in the publication of 1931 he made two suggestions which I condense in two quotations:

> The ectoderm of the fully formed gastrula could possess some kind of stratification or gradient from the cephalic to the caudal pole, whereby the regional determination of the primary as well as that of the secondary neural plate would be affected. (Spemann, 1931a, p. 503)

> The alternative explanation deserves serious consideration: that the primary organization field exerts a determining influence on the secondary one. . . . It is possible that the induced anlage receives its regional pattern under the influence of the primary anlage, so-to-speak as its mirror image. (Spemann, 1931a, p. 507)

In other words, he envisages that the axial organs of the host embryo create (overlapping) induction fields that extend to the ventral regions much like those illustrated above in Fig. 6–2c. They would then exert a region-specific influence on inductions evoked in the ventral ectoderm. But since the implanted normal or abnormal inductors possess their own intrinsic region-specificity, mutual interactions would occur and one or the other factor would become dominant and decide the outcome.

The problem of the host influence on regional specificity has remained elusive. I have dealt with it to set forth once more the immense complexity of animal development and the painful inadequacy of the methods then available to solve even those problems which were clearly defined. But defining them was perhaps an important step. From our vantage point, the results of Chuang's experiments are reassuring. While they demonstrate a distinct influence of the host embryo on the regional specificity of inductions, the effect of the host is not sufficiently strong to discredit the Einsteck-method as a tool for the analysis of the activity of abnormal inductors.

Chuang's experiments firmly established the notion that there exist at least two inductive agents with region-specific effects: a clearly defined archencephalic agent, which induces forebrain, eyes, and balancers, and a less clearly defined agent, which induces trunk and tail structures, including spinal cord, notochord, somites, kidney tubules, and fins. It may be called provisionally the spinocaudal agent. The experiments have confirmed the heat-stability of the former and the heat-lability of the latter. The origin of the deuterencephalic structures—the midbrain, the hindbrain, and the adjacent ear vesicles—has remained in doubt.

Toivonen's experiments, like those of Chuang, took as their point of departure Holtfreter's finding that adult animal tissues induce a great variety of structures in gastrula ectoderm and that they fall into two groups: neuralizing and mesodermizing inductors. But Toivonen was interested primarily in effect-specificity (Leistungs-Specifität), that is, in the *type* of inductions which are evoked, rather than in the influence of the different host levels on the outcome. We shall see that his shift of emphasis moved the analysis a great step forward.

Toivonen tested over forty vertebrate tissues for their inductive capacity and eventually chose eight which showed the most distinct effect-specificity: the livers of the perch *(Perca fluviatilis),* the viper *(Vipera berus),* and the guinea pig *(Cavia cobaya);* the kidneys of the perch, the viper, the guinea pig, and the jay *(Perisoreus infaustus);* and the thymus of the guinea pig. I shall omit the thymus experiments, because they yielded only a low percentage of isolated lenses and fragments of neural tissue. All tissues were pretreated with 70% alcohol for several hours or weeks, washed, and inserted in the blastocoele of gastrulae of *Triturus taeniatus* (Einsteck-method). The experiments were done on a sufficiently large scale to permit a statistical treatment of the data.

An essential ingredient of Toivonen's success was the choice of categories which he used to classify the inductions. He divided them into four groups: (1) *anterior head structures,* including telencephalon, diencephalon, eyes, noses, and balancers; (2) *posterior head structures,* including mesencephalon, rhombencephalon, and ear vesicles; (3) *trunk structures,* including spinal cord, notochord, somitic musculature, kidney tubules, and dorsal fin; and (4) *tail structures,* including the axial organs and tail fin. It will become apparent later that Toivonen's most important advance was to subdivide Spemann's head organizer into anterior and posterior head inductors, referred to as *archencephalic* and *deuterencephalic.* This idea had suggested itself by the outcome of the experiments. The three liver preparations gave a strikingly uniform picture: they induced almost exclusively archencephalic structures and a small percentage of deuterencephalic structures. In contrast, the kidneys induced rather complex combinations in which trunk and tail structures prevailed (Fig. 9–4). It should be mentioned that the identification of archencephalic inductions was based largely on the presence of eyes, noses, and balancers, since the brain vesicles were frequently atypical and not identifiable as telencephalon or diencephalon. Likewise, the posterior head level could be identified by the ear vesicles, which hardly ever failed to differentiate, even when the hindbrain was poorly developed.

The influence of different host levels on the type of inductions was not ignored by Toivonen, but, on the contrary, it was discussed in detail. On the basis of the statistical analysis of his data he came to the same conclusion as Chuang: that the host influence is clearly demonstrable, but it is expressed only in the quantitative distribution of the different types of induction. Archencephalic inductions are much more frequent at the head and heart levels than at the trunk level, and the reverse is true for trunk and tail inductions. The effect-specificity of the inductors is never overruled by the host influence, though it can be modified by it. In summary, one can state that there are indisputably differences in the regional character of the inductions which depend on the host embryo, even though they reveal themselves only in a very general way, as, for instance, in the gradual decline of

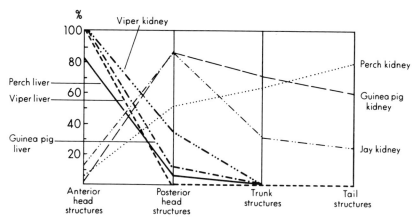

Fig. 9–4. Graphic representation of the induction-specificity of seven adult tissues indicated on the graph. The tissues were pretreated with alcohol. The region-specific types of induced structures are indicated on the abscissa. The frequency of the induced tissues (in percent) is indicated on the ordinate. From Toivonen, 1940.

the head structures and a corresponding increase of trunk-tail structures in craniocaudal direction (Toivonen, 1940, p. 106). Thus the emphasis is shifted to the properties of the abnormal inductors. "The only remaining cause of the difference in the effects is that the inductors contain different effect-specific inductive substances, or [each contains] several in varying ratios" (Toivonen, 1940, pp. 120–121).

Before I pursue this topic further, I call attention to a particularly interesting case. An implant of perch kidney had produced a protrusion from which a well-developed tail had grown out. The microscopic sections showed, in addition, a hindbrain flanked by a pair of ear vesicles. The occurrence of head structures and tails, side by side, observed also by Chuang, makes an important point: if in normal development heads and tails originate at opposite ends of the main axis, this means that additional factors operate to position the specific inductions in an appropriate topographic pattern along the main axis. In other words, the inductions have to acquire "positional information" in addition to regional specification. Those *abnormal* inductors which contain several different inductive agents are not under this constraint; they can release them at the same spot.

Once the existence of several effect-specific agents was established, the next step was to inquire how many inductive substances have to be postulated to account for the experimental results. A glance at Fig. 9–4 shows that one group of tissues, the livers of perch, viper, and guinea pig, stands apart from another group, the kidneys of perch, jay, and guinea pig, while the kidney of the viper takes an intermediate position. As was mentioned, the livers induced almost exclusively archencephalic structures. The remarkable uniformity of the forebrain inductions strongly supports Chuang's supposition of an archencephalic inductive agent.

The inductions by kidneys include all structures of the main axis from the posterior head to the tail. They present greater difficulties for the analysis. Although these structures frequently occur together as a single package, Toivonen expresses legitimate doubts that only one specific inductor is involved. What seems reasonable to infer, however, is that a *mesodermizing* component plays a major role:

> Since in normal development the spinal cord is induced by the underlying chordamesoderm, the most natural explanation for the origin of a regular trunk and posterior body is the assumption that the heterogenous [abnormal] inductors of this group contain an inductive agent which causes primarily the formation of the mesoderm. This in turn calls forth secondary neural structure [the spinal cord] and this finally initiates, as the third step, the formation of the dorsal fin. (Toivonen, 1940, p. 125)

This factor corresponds to Holtfreter's mesodermizing agent. In common with the latter, it is inactivated by heat, as was established by Toivonen in a later experiment (1949).

While the evidence for an archencephalic and a mesodermal agent is rather compelling, the case for a third specific agent which induces hindbrain and ear vesicles is not very strong. Toivonen advances several arguments in its favor, which I shall not discuss in detail. The explanation of the experimental results in terms of three specific inductive agents is certainly plausible. However, Lehmann offered a very intriguing alternative in the aforementioned review article (1942). After the discussion of the results of Chuang and Toivonen and the acknowledgment of the existence of a heat-stable archencephalic and a heat-labile spinocaudal factor, he says:

> It is striking that the deuterencephalic factor is weakened by long-lasting heat. One will have to find out whether this factor is represented by a certain quantitative ratio of the archencephalic and the spinocaudal factor, hence that it does not represent a special substance at all. Furthermore, the fact that the deuterencephalic zone is a transitional region between the archencephalic and the spinocaudal region would speak in favor [of the notion] that the deuterencephalic induction effect is brought about by a mixture of the archencephalic factor and the spinocaudal factor" (Lehmann, 1942, p. 524).

Indeed, Lehmann's interpretation turned out to be the correct one. But he was far ahead of his time. It was not until 1955 that Toivonen and Saxén, inspired by Lehmann's suggestion, performed an experiment that supported his view.

Lehmann was a Swiss experimental embryologist who had been a student of Baltzer in Bern. Much later he became Baltzer's successor as the director of the Zoological Institute of the University of Bern. He spent several years in Spemann's laboratory, where I got to know him well. As the quoted article shows, he had a strong theoretical bent; this also finds expression in his textbook (Lehmann, 1945). He was less innovative in his experiments. If he had tried, and been successful in providing experimental evidence for his intuition, he would have been better remembered than he actually is.

A final comment on these pioneering studies. At first sight, Toivonen's experiments could give the impression that particular animal tissues are the carriers of specific inductive agents. But the viper kidney behaves differently from the other kidneys, and Chuang had found that the salamander liver, in contrast to those used by Toivonen, also induces trunk structures and tails. Later experiments confirmed that there is no correlation whatsoever between specific animal tissues and their effect-specificity as inductors.

When one speaks of region-specific inductors, one should be aware that the "instruction" which they give to the reacting tissue is of a very general nature. For instance, the archencephalic agent induces a forebrain-eye field which then proceeds to differentiate further by self-organizaiton. The main actor is not the inducing agent but the remarkably pluripotent gastrula ectoderm with its capacity for the production of a large spectrum of embryonic structures and their patterned organization.

Double-Gradient Theories of the Origin of Axial Organization (1950–68)

The conceptual framework

Little progress was made in the 1940s, due in large measure to the war and its aftermath. In the early 1950s, the analysis of embryonic induction by means of abnormal inductors was resumed with renewed vigor. (Both Chuang and Toivonen had used the term "heterogenous" synonymously with "abnormal." I shall use the former term henceforth.) The experiments followed two different paths. The biochemically inclined embryologists used the then available methods for the fractionation and purification of the inductive agents. Others continued the analysis of the origin of regional specificity, using vertebrate tissues as heterogenous inductors. I shall limit myself to the latter issue.

I have quoted above the seminal idea of Lehmann (1942) that the sequence of regional differentiations along the main axis could be explained by assuming that the anterior head is induced by an archencephalic agent, the posterior trunk and tail by a spinocaudal agent, and the intermediate regions by appropriate mixtures of the two agents. Later, in a resumé of a symposium, he formulated his idea in semiquantitative terms:

> The findings discussed in the symposium suggested to me the assumption that in a neural functional state [Funktionszustand] there are always present two antagonistic factors, an archencephalic and a spinocaudal [factor] which are polarized . . . and overlap like the animal and the vegetal functional states in the sea urchin embryo. The ratio of the mixtures of the two factors could determine the definitive regionality: much archencephalic + little spinocaudal = archencephalic effect; moderate archencephalic + moderate spinal = deuterencephalic effect; little archencephalic + much spinal = spinal effect. (Lehmann, 1950, p. 144)

The quote requires several comments. The term "functional state" is explained elsewhere as the equivalent of "capacity for a particular differentiation" (Lehmann, 1937, p. 4. Lehmann was always fond of florid terminology). The term "antagonistic" is explained as follows: "It seems that the two factors are antagonistic; in the live inductors used by Chuang, the spinal factor seems to prevail over the archencephalic factor. But as soon as the weakening of the spinocaudal factor increases, the effect of the archencephalic factor becomes stronger" (Lehmann, 1942, p. 524).

The reference to the sea urchin embryo is very significant. In the 1930s, the Swedish experimental embryologists J. Runnström and S. Hörstadius had proposed a double-gradient theory to explain the differentiations along the main axis which connects the animal (upper) and the vegetal (lower) pole of the egg. They assumed an "animal" gradient *(an)* with its peak at the animal pole and a "vegetal" gradient *(veg)* with its peak at the vegetal pole. According to the theory, the development of the different structures along the axis is determined by appropriate ratios of the two agents. For instance, a high *an/veg* ratio causes the formation of a tuft of long cilia, the apical tuft, at the animal pole of the larva, and a high *veg/an* ratio causes invagination and skeleton formation by the vegetal zone. Crucial evidence for the theory was provided by extensive, ingenious experiments of Hörstadius. He managed to isolate and recombine cell layers of different cleavage stages in all possible permutations, to manipulate the tiny eggs in other ways, and to rear the experimental animals to larval stages. This was an extraordinary feat. I mention one experiment to illustrate the notion of gradients. If the *an/veg* ratio is shifted towards *an* by the removal of cells at the vegetal pole, then the apical tuft becomes enlarged, and the extent of the area covered by cilia is proportional to the amount of vegetal material that was removed (review in Hörstadius, 1939, 1973).

I have not dwelt on the gradient theory because Spemann was critical of it and not particularly interested in it. (His criticism is discussed in a chapter in his book.) Yet the concept of gradients received considerable attention in the 1920s and 1930s, due largely to the initiative of its major proponent, the American developmental physiologist C. M. Child of the University of Chicago (Child, 1941). The basic tenet of his axial gradient theory was the assertion that the structural organization along the main axis of organisms is causally related to underlying gradients of metabolic activity. Child derived his experimental data mainly from regeneration experiments. A frequently cited example was taken from the regeneration of fragments of the flatworm *Planaria*. The anterior cut surface would regenerate a head, because it was the site of the highest metabolic activity referred to as the "dominant region"; the posterior cut surface, representing the lowest metabolic activity, would regenerate the tail. The theory was also applied to embryos. The upper blastoporal lip of the gastrula was supposed to be the site of the highest respiratory rate, hence the dominant region. The properties of the organizer were derived from its preferential metabolic status (Child, 1929, 1946). The major appeal of the gradient theory was its reductionist approach to an understanding of axial organization, though Child never lost sight of the organism as a whole. But many—Spemann among them—found it difficult to envisage how continuous metabolic gradients could explain the discontinuity of structural patterns. (Here, a stronger emphasis on thresholds might have helped.) Others took a middle ground. They rejected Child's emphasis on *metabolic* gradients and preferred to think in terms of concentration gradients of chemical agents. The double-gradient theory applied to sea urchin development is an example. The same idea of double gradients provided the theoretical framework for the experiments to be discussed below.

Lehmann's idea of the role of two opposing gradients in the determination of region-specificity of the neural axis was an inspired guess. T. Yamada was the first to formulate a substantive double-gradient theory based on his own experimental work and that of others (Yamada, 1950). He states his basic assumptions as follows:

> 1) The developmental activity of a given germ region depends . . . on two sorts of 'potentials' which are designated respectively as dorso-ventral [dv] and cephalo-caudal [cc]; they can vary relatively independently of each other. 2) At a given developmental stage, the dv and cc potentials show a definite value along the respective axis; or, expressed differently, every germ region has a definite value of both potentials according to its topographic position relative to both axes. 3) In the course of development, the values of both potentials show a continual change which is characteristic for the germ region and for the stage (Yamada, 1950, pp. 13–14).

Peak values of the dorsalizing (dv) potential are associated with the differentiation of brain and spinal cord. The cc potential has its highest value in the caudal region and decreases toward the cephalic end of the main axis. A region with very high dv potential and very low cc potential will differentiate to forebrain and eyes; a decrease of the dv value in the same region will result in the differentiation of lens and nose, and further decrease in the formation of mesenchyme and melanophores. Caudal structures, such as notochord, posterior somites, and tail are differentiated when the cc potential is relatively high and the dv potential is relatively low. The formation of the deuterencephalon (hindbrain) requires a high dv level and an intermediate cc level.

According to the theory, the dv and cc potentials are activated by two distinctly different mediators or inductors, Mdv and Mcc. "The dorso-ventral potential is assumed to be connected primarily with the intensity of some of the biochemical activities of the developing system, while the cephalo-caudal potential is connected primarily with the morphogenetic movements of the developing system" (Yamada, 1950, p. 14). The latter statement requires an explanation. An intriguing and novel aspect of Yamada's theory is the idea that morphogenetic movements play a decisive role in the determination of the fate of embryonic parts. One is reminded of Vogt's distinction between dynamic and material determination. According to the theory, the cc gradient is caused not by a biochemical agent but by differences in the degree of particular morphogenetic movements. The critical movements during axis determination are the gastrulation and neurulation movements. They have been analyzed in detail in the classical studies of Vogt and his student, Goerttler, using the vital-staining method. I shall deal specifically with the neurulation movements. While they occur, the animal pole region, which is the prospective forebrain area, remains stationary. The more caudal regions of the incipient neural plate undergo elongation by stretching, and the more lateral regions are at the same time compressed toward the midline by a process called *convergence*. "The more caudal the germ region the more intensive the stretching-convergence" (Yamada, 1950, p. 15; see also Fig. 2–11). For instance, the spinal cord in the tail would result from a high level of both the cc and the dv potentials.

A crucial point in Yamada's theory is the assumption that the gradient of patterned cell movements in the neural plate is not caused by region-specific biochemical properties, but by a corresponding pattern of dynamic properties of the underlying chordamesoderm (the organizer). It is a dynamic process of morphogenetic movements in the mesoderm which induces the cc gradient in the overlying ectoderm. In other words, both biochemical and dynamic processes cooperate and are integrated in the creation of region-specificity, not only in the brain and spinal cord but in other structures as well (balancers, suckers). A final point: if the dynamic properties of the chorda-mesoderm are not related to a chemical agent, to what other property can they be attributed? According to the theory, it is a *structural* element in the organizer which endows it with its dynamic region-specificity. If the structure of the organizer is destroyed, for instance by heating, then it loses its capacity to perform and to induce morphogenetic movements; it retains only the biochemical agent, or dv mediator, which enables it to induce archencephalic structures, that is, forebrain, eyes, and others. It will be shown below that Yamada's basic scheme of a double-gradient theory was adopted later by Toivonen and Saxén, but with the important modification that both the dorsoventral and the cephalocaudal gradient were attributed to biochemical inductive agents.

A mesoderm-inducing agent

In the experiments discussed so far, the inductions of mesodermal trunk and tail structures had always been accompanied by a spinal cord; hence the choice of the term "spinocaudal." The origin of the spinal cord could be explained in two ways: that it was induced secondarily by the mesodermal structures, by analogy with neural induction in the organizer experiment, or that the heterogenous inductors contained both a neuralizing and a mesodermizing agent. The latter hypothesis could not be tested because a pure mesodermizing agent was not available. Toivonen's (1953a,b) announcement that he had found in the bone marrow of the guinea pig a mesoderm-inducing tissue was therefore received with great interest. His attention had been directed to the bone marrow by reports of several investigators that extracts of bone marrow when injected into musculature would "induce" cartilage and bone formation; and it had been suggested that the agent might be related to the heterogenous inducing agents (for references see Toivonen, 1953a).

In Toivonen's experiment, the bone marrow was pretreated with alcohol and brought in contact with gastrula ectoderm by the Einsteck-method (Toivonen, 1953a,b) and by the sandwich method (Toivonen, 1954). In both instances, the bone marrow induced only mesodermal derivatives (notochord, somitic musculature, kidney tubules), but no neural structures. When it was placed in water heated to 80–90°C for 10 minutes, the mesoderm-inducing agent was inactivated, and only a few isolated lenses and balancers were induced, indicating the admixture of a weak archencephalic agent. Toivonen concluded that "the experiments show unequivocally that the mesoderm-inducing agent is entirely independent of the agents inducing neural structures" (Toivonen, 1953a, p. 99). Later Yamada repeated the bone marrow experiment on a larger scale, as a control to an interesting experiment of his own. Using the sandwich technique, he confirmed Toivonen's results. In fact, the inductions he obtained were better organized than those in Toivonen's experiments; typical mesodermal axial organs, including a notochord and two adjacent rows of somites, occurred in a high percentage of cases. The absence of the spinal cord was again conspicuous (Yamada, 1959). While all these experiments leave no doubt about the presence of a strong mesodermizing agent in the bone marrow, the failure of notochord and somites to express their ability for neural induction, which had been taken for granted since the organizer experiment, is a puzzle which to my knowledge has not been resolved.

Experimental tests of two-gradient theories

In the early 1950s, Toivonen established a partnership with his colleague, L. Saxén, which left its mark on experimental embryology to this day. Their first collaborative effort was a test of the hypothesis that the entire range of region-specific structures, from head to tail, can be explained by the combined activities of an archencephalic and a mesodermizing agent. They used the guinea pig liver as the archencephalic inductor and the guinea pig bone marrow as the mesodermizing inductor. They were aware that the bone marrow contained a small admixture of a neuralizing agent. Small fragments of both tissues were implanted side by side in the blastocoele or in an ectodermal envelope (Fig. 9–5; Toivonen and Saxén, 1955a,b). In both experiments, the inductions formed solid masses which ended in typical tails. The microscope sections revealed that a fairly complete assortment of mesodermal and neural structures had been induced. The results are tabulated in Fig. 9–6. The most significant finding was the appearance of spinal cords in almost 100% of the cases and of deuterencephalic structures (hindbrain, ear vesicles) in 60% of the cases ($n = 66$). The authors concluded that since the spinal cord was never induced by the liver alone and only rarely by the bone marrow alone, its regular presence in the combination experiment must be due to the joint activity of the two inductors. Furthermore, the percentage of hindbrain inductions was greatly increased over that produced by the liver alone. Figure 9–6 shows that, indeed, the entire spectrum of embryonic differentiations had been induced by the two tissues.

The Einsteck-experiment delivered even more than expected. In a number of cases, the induced structures were arranged in their normal sequence along the main axis, simulating a secondary embryo. The authors suggest that each inductor projects a determination field onto the ectoderm, and that the deuterencephalic structures and the spinal cords were induced in the region where the gradient fields overlap (Fig. 9–5). The simulation of the organizer experiment by the combined activities of two fragments of mammalian tissues is indeed a remarkable feat. The credit should be shared by the investigators for the design of the experiment and the judicious choice of the heterogenous inductors, and by the gastrula ectoderm which displayed once more its uncanny capacity to decode anomalous signals and to respond to them in a meaningful way.

Although the experiment was qualitative in nature, it invited an interpretation in quantitative terms. This took the form of a double-gradient theory, as represented in Fig. 9–7. One notices that the mesodermizing agent (M) is distributed in a gradient fashion along the main axis, whereas the neuralizing agent (N) is evenly distributed along the anterior-posterior axis. However, the N-agent has a

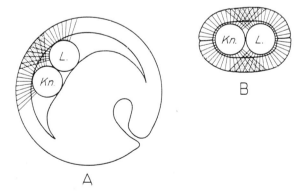

Fig. 9–5. A. Implantation of a neuralizing inductor (liver of guinea pig, L) adjacent to a mesodermizing inductor (bone marrow of guinea pig, Kn) into blastocoele of early gastrula of *Triturus taeniatus*. B. Implantation of the same inductors into ectoderm vesicle (sandwich experiment). From Toivonen and Saxén, 1955b.

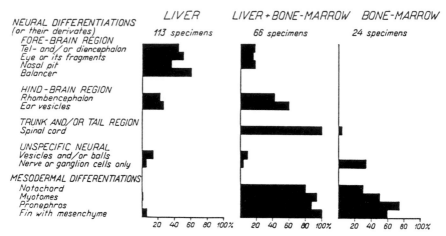

Fig. 9–6. Diagram of the frequency of inductions of various tissues by guinea pig liver and bone marrow, implanted separately or combined into the blastocoele of gastrulae of *Triturus taeniatus.* From Toivonen and Saxén, 1955a.

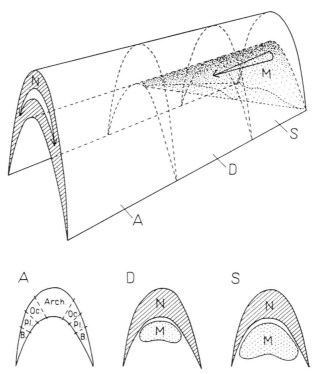

Fig. 9–7. Diagram of Toivonen and Saxén's double-gradient theory of inductions by the combination of a neuralizing agent (N) and a mesodermizing agent (M). A, Archencephalic region, D, deuterencephalic region; S, Spinocaudal region. The dorsoventral gradient of the neuralizing agent at the archencephalic level is shown in (A) in the lower row. Oc, Eye; Pl, placode; B, balancer. From Toivonen and Saxén, 1955b.

graded distribution in the mediolateral direction. M extends forward only to the deuterencephalic level (D). At its highest concentration in the dorsal midline, N induces the different parts of the central nervous system. When reduced in strength in ventral direction, it induces sense organs (eye, ear vesicles): a further weakening results in the induction of placodes (nose, isolated lenses), and its weakest expression is the induction of balancers (see A in the lower row of Fig. 9–7). The authors concede that the N-agent may also be graded along the main axis, but their material did not provide support for this assertion. We shall see that in its final version, the double-gradient theory postulates two opposing gradients along the main axis.

The next goal was to quantify the mixtures of the two agents. Since purified fractions were not available at that time, the investigators designed an experiment in which cells containing different inductive agents were mixed in different ratios. Their choice was HeLa cells, human tumor cells that had been grown in tissue culture in human serum for decades. It had been found earlier that untreated HeLa cells are strong spinocaudal inductors with a weak deuterencephalon-inducing component, whereas heat-treated HeLa cells are pure archencephalic inductors. The HeLa cells were centrifuged, and the sediment was shaken to form a homogeneous mass, which was then divided in two equal halves. Samples of the two groups were counted. One half was heated at 70°C for 10 minutes. The two suspensions were then mixed at the following ratios: 9:1, 7:3, 1:1, 3:7, and 1:9. Great care was taken to keep the mixtures dispersed homogeneously. The mixtures were then centrifuged again and the compact cell masses were tested using the Einsteck-method (Saxén and Toivonen, 1961). The results are represented in Fig. 9–8. One notices that the untreated cells are not pure mesodermal inductors; they also induce spinal cord and even a low percentage of deuterencephalic structures. This is a distinct shortcoming; nevertheless, the spectrum of the structures induced by the combined action is markedly different from that induced by each component alone.

> In mixing these two inductors with different regional inductive actions, a progressive shift of the regional character of the inductions was observed, obviously as a result of the combined action of the components. Starting from the purely archencephalic-inducing heated cells, even an addition of 10% of non-heated cells resulted in an induction of hindbrain structures to the extent of almost 100%. . . . Thus the inductive action of this mixture is definitely a *new* one and does not represent the regional inductive types of either of the two components of the mixture. (Saxén and Toivonen, 1961, pp. 522–523).

An addition of 30% of nonheated cells leads to the induction of a complete set of all regional types, with a preponderance of deuterencephalic structures. It is clear that the stepwise shift in the ratios of the mixtures is paralleled by the shift of region-specific inductions from archencephalic to caudal. Since heating does not affect the archencephalic agent, its quantity remains constant in all mixtures. Hence, what the experiment proves is the existence of a single gradient of the spinocaudal agent, produced by the unheated cells. In this respect, the results reflect exactly the schema illustrated in Fig. 9–7. Despite its shortcomings, the

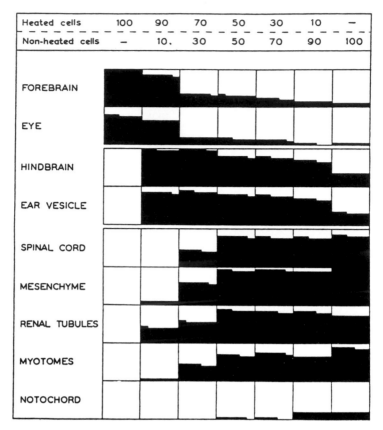

Heated cells	100	90	70	50	30	10	—
Non-heated cells	—	10.	30	50	70	90	100

FOREBRAIN

EYE

HINDBRAIN

EAR VESICLE

SPINAL CORD

MESENCHYME

RENAL TUBULES

MYOTOMES

NOTOCHORD

Fig. 9–8. Percentage of inductions of different tissues by a graded series of mixtures of untreated HeLa cells and HeLa cells killed by heat treatment. From Saxén and Toivonen, 1962.

experiment is of historical interest as an intermediate step between qualitative and quantitative tests of the double-gradient theory.

In the 1950s, a consensus had been reached as to the chemical identity of the inducing agents. The active ingredients of both the neuralizing and the mesodermizing tissues had been characterized as proteins. The biochemical pioneering work had been done by Yamada and his associates (Yamada, 1950, 1961) and by Toivonen (1949). They were joined in the mid-1950s by a German group led by Heinz Tiedemann. He and his wife Hildegard Tiedemann had been associated with O. Mangold in Heiligenberg. In 1967 he moved to the free University of Berlin, where he is now professor of molecular biology and biochemistry.

Tiedemann departed from the tradition of using mammalian tissues as a source of inducing agents and turned to 9-day chick embryos. His fractionations led to the isolation of purified neuralizing and mesodermizing agents. A variety of fractionation procedures were developed (Tiedemann and Tiedemann, 1959; Tiedmann 1963, 1967, 1975, 1978). Briefly, the heads were removed to get rid of large amounts of lipids, and the bodies were processed as a whole. Purified fractions were suspended in water, homogenized, and precipitated in alcohol. They were then implanted in gastrulae (Einsteck-method). Different fractions induced a wide spectrum of region-specific structures. A highly purified fraction which induced only mesodermal structures was identified as a protein with a molecular weight of 25,000–30,000. A less highly purified nucleoprotein induced only archencephalic structures. Treatment with proteolytic enzymes and ribonuclease showed that the protein moiety was the active component. In later studies, highly purified mesoderm-inducing proteins were found also to induce endodermal structures, and the term "vegetalizing agent" was coined to include both. However, the experiments to be discussed below deal only with pure mesodermizing fractions, and I shall continue to use this term.

Quantitative tests of the two-gradient theory could now be carried a step further by using mixtures of proteins rather than mixtures of cells. Highly purified archencephalic and mesodermizing proteins were mixed in ratios of 1:1, 5:1, and 50:1 (Tiedemann, 1963; Tiedemann and Ticdemann, 1964). The archencephalic fraction alone induced large forebrains, eyes, and noses and a small percentage of deuterencephalic structures. The mesodermizing fraction alone induced large amounts of muscle tissue and kidney tubules and, less frequently, notochords. The results of the combination experiments are shown in Fig. 9–9. As in the preceding experiments, the mixtures induced structures which were not induced by either one of the components alone, such as spinal cords and tails. The mixture 5:1 is of special interest in that it produced almost the entire spectrum of axial structures. In several cases, the inductions represented a typical axial system from the hindbrain-ear level to the tail. There are no illustrations of these embryos, but they must have resembled the famous secondary embryo *Um 132* in H. Mangold's organizer experiment, which also extended from the hindbrain to the tail. As expected, the mixture 50:1 shifted the induction types further toward the anterior head region, while trunk and tail inductions disappeared almost completely. "From these results it may be concluded that a small amount of neural factor combined with a large amount of the mesodermal factor is needed to obtain complete tails with neural tubes, and that, on the other hand a rather large amount of the neural factor together with a smaller quantity of the mesodermal factor is necessary for the induction of hindheads" (Tiedemann, 1963, p. 197). This is an almost verbatim rendition of Lehmann's proposition of 1950. But Tiedemann was aware of the limitations of the experiment. "Our experiments give no information as to the real concentration ratios of the pure factors which result in the deuterencephalic and spinocaudal inductions. They were carried out not with pure factors but with more or less concentrated neural or mesodermal inducing fractions" (Tiedemann and Tiedemann, 1964, p. 132).

This combination experiment supplements the experiment with HeLa cells in an important point. In the latter, the quantity of the neuralizing agent had been kept constant and that of the mesodermizing agent had been increased stepwise. The reverse occurred in the mixture of the proteins. Taken together, the experiments provide a sound foundation for the double-gradient theory.

The title of this section implies that there is more than one double-gradient theory. This point needs an explanation. The original version (Fig. 9–7) postu-

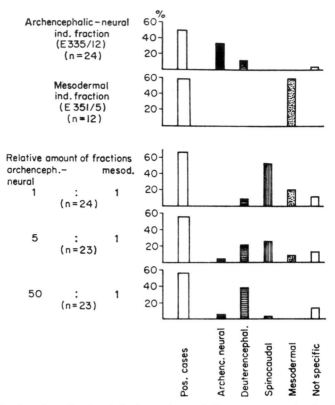

Fig. 9–9. Inductions by a fraction inducing archencephalic structures and a fraction inducing mesodermal structures and their combination in ratios 1:1, 5:1, and 50:1. From Tiedemann, 1963.

lated an axial gradient of the mesodermizing agent in the posterior-anterior direction and a dorsoventral gradient of the neuralizing agent. Both the HeLa experiment and the combination experiment with two proteins assume two gradients along the main axis, in opposite directions. Hence we are dealing actually with three gradients. In the following discussion double-gradient theory refers to two axial gradients with their peaks at opposite ends of the main axis.

Application of the double-gradient theory to normal development

The performers of the experiments with heterogenous inductors realized that their efforts were ultimately directed toward an understanding of the regional inductor-specificity of the organizer in normal development. They were in fact in pursuit of a deeper analysis of Spemann's head and trunk organizer. But they stated their goal only in very general terms. "It seems that our two-gradient hypothesis . . . provides a satisfactory explanation of the regionality caused by the action system in the primary induction process" (Toivonen, 1961, p. 87). But on closer inspection, the situation does not appear to be that simple; it is actually rather problematical. There seems to be no difficulty in equating the archencephalic agent with the head organizer or, more specifically, with the inductor of the forebrain-eye field. But the role of the mesodermizing agent in normal development is not immediately obvious. Two separate issues come to mind. First, the two-gradient hypothesis postulates that the posterior end of the invaginated organizer or archenteron roof induces mesodermal structures. Yet traditionally the entire neural plate was considered to be the precursor of the nervous system, as its name implies. Second, the organizer transplantation experiment had clearly shown that the organizer itself, that is, the chordamesoderm region, is determined to its fate at the early gastrula stage; therefore, a mesodermizing agent would have no meaningful role to play, except perhaps to stabilize the chordamesoderm determination. Of course, one can ask how the chordamesoderm had acquired its self-differentiation capacity to begin with, and one can look for a mesoderm-induction process in pregastrulation stages. A series of very interesting experiments in the last decades has provided evidence that, indeed, the determination of the mesoderm results from an inductive interaction between the ventral endoderm cells and the prospective ectoderm cells located above the equator. This process occurs in late cleavage and blastula stages. I cannot go into the details; this would detract from my major theme. I refer to the reviews by Nieuwkoop (1973), Nakamura (1978), Gerhart (1980), and Gimlich (1985).

The other question—whether the prediction of the double-gradient theory that the posterior end of the mesoderm mantle or archenteron roof induces mesoderm in the overlying ectoderm is valid—has also found a satisfactory solution in the affirmative. It may come as a surprise to embryologists who are not familiar with the intricacies of neurulation (the transformation of the neural plate to the neural tube) in amphibians that indeed the posteriormost one-fifth of the neural plate forms not spinal cord but the posterior trunk somites and two rows of tail somites, all of which are mesodermal derivatives. This information was provided first by the Dutch experimental embryologist Hubertha Bijtel, a student of Woerdeman (Bijtel, 1931). Using the axolotl, *Ambystoma mexicanum*, she placed small vital-staining marks at the posterior end of the neural plate at different stages of neurulation, and she followed the markings, which can be observed through the skin, to larval stages (Fig. 9–10b,c). First, she confirmed what Vogt had observed in his vital-staining experiments: that invagination continues for a while after gastrulation is completed and the blastopore has become a slit. Bijtel located the material which had invaginated late in the posterior somites of the trunk. But this process soon comes to an end; it is followed by a conspicuous stretching process of the posterior end of the neurula whereby the tail is formed. Markings placed at the posterior one-fifth of the neural plate—a very small area—were found eventually to stain both rows of the numerous tail somites. The small area had expanded and stretched enormously. But, strangely enough, the notochord and the spinal cord of the tail had consistently remained unstained. Where is the material for these axial structures located in the neurula? This question was answered by vital-staining the fourth fifth of the neural plate. It turns out that this small region likewise stretches and rapidly grows backwards. It catches up, so to speak, with the out-growing posterior one-fifth. The notochord material for posterior trunk and tail, in growing backward, places itself between the two rows of somites. The spinal

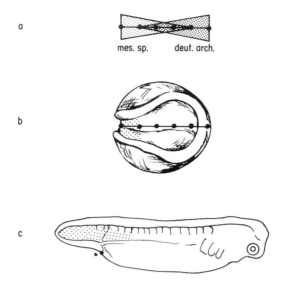

Fig. 9–10. a. Schema of a double-gradient theory. Distribution of neuralizing *(dense stipple)* and mesodermizing inductive agent *(light stipple)* in archenteron roof of late gastrula. arch., Archencephalon; deut., deuterencephalon; sp, spinal-trunk; mes., mesoderm-inducing region. Original drawing. b. Vital-staining of posterior one-fifth of neural plate. c. Same embryo in larval stage. Posterior trunk and tail somites are vital-stained. b and c from Bijtel, 1931, modified.

cord material which was located in the neurula in front of the prospective notochord material keeps pace with the elongation of the other axial organs and stays in the dorsal midline. When the neural folds adjacent to the posterior neural plate were stained, the markings were found in the tail fins. This very precise account of the derivation of the tail structures was confirmed later by others. It dispelled once and for all the then prevailing view that the structures of the tail originate from a mass of indifferent cells, the tail bud.

What was required next was experimental evidence that the mesodermal posterior fifth of the neural plate had been induced by the underlying posterior end of the invaginated organizer. The proof was provided by transplantation experiments performed by the American experimental embryologist Walter R. Spofford, a student of Harrison (Spofford, 1948). He removed the posterior end of the neural plate in early neurulae of the salamander, *Ambystoma punctatum,* and replaced it with ectoderm of an early gastrula which had been vital-stained before the removal of the transplant. As expected, the transplant formed two rows of posterior trunk and tail somites, while the notochord and spinal cord were unstained; they derived from material located in front of the transplant. When the transplants were made a bit larger and extended into the fourth fifth of the neural plate, then notochord and spinal cord were also stained. A major premise of the double-gradient theory was thus validated: the posteriormost region of the organizer induces mesodermal structures.

It remained to be shown that the region of the neural plate between the forebrain-eye field and the tail somites, that is, the prospective hindbrain and spinal cord, acquire their regional specification by the combination of different ratios of the mesodermizing and the neuralizing (archencephalic) agent. This challenge was difficult to meet. Nobody assumed that the two protein fractions isolated by the Tiedemann team would be identical with the agents produced by the organizer in normal development; hence using combinations of these two fractions would not prove the point. Nor was it technically feasible to obtain sufficient amounts of embryonic material for fractionation. The only recourse was to turn again to mixtures of cells. This step was taken by Toivonen and Saxén (1968; see also Toivonen, 1978). Very early neurulae of the common European newt, *Triturus vulgaris,* were used. A piece of the anterior neural plate which is an archencephalic inductor (O. Mangold, 1933) and a piece of the posterior trunk region of the organizer were removed and the fragments combined in ratios of neuralizing and mesodermizing cells of 10:1, 5:1, 5:2, 5:5, 2:5, and 1:5. The combined tissues were disaggregated and very thoroughly mixed. They were then grown in small depressions in the agar bottom of the culture dishes for 14 days and studied microscopically. The results, presented in percentage of structures differentiated, are unequivocal (Fig. 9–11). As long as the neuralizing cells prevail (10:1; 5:1), only archencephalic structures from the anterior cells and mesodermal structures from the posterior cells are differentiated. This is the result of the self-differentiation of the starting materials. At the ratio 5:2, the first hindbrain and ear vesicles were observed at a frequency of 40%. Their frequency rose to 100% when equal parts of the neuralizing and mesodermizing cells were mixed. This mixture produced also the first spinal cord differentiations; their percentage rose with the prevalence of the mesodermizing agent at ratios 2:5 and 1:5. Clearly, the hindbrain and spinal cord differentiations resulted from the combined effects of the two inductors, and their frequency shifted with the stepwise shift of the ratios, as predicted by the theory.

This experiment is the best evidence we have so far of the validity of the double-gradient theory in normal development. It falls short of a test with cell-free extracts of the two cell types, but nonetheless it leaves no doubt that a neuralizing and a mesodermizing inductive agent operate in the amphibian gastrula and neurula in accordance with the double-gradient theory.

The authors draw another conclusion: that the region-specific determination of the different parts of the neural plate is a two-step process. "During the initial stage of induction the cells are determined to become neural, but they acquire no stable regional character. This is subsequently controlled by the mesodermal cells and apparently in a quantitative way, since an increasing amount of mesoderm surrounding the neural cells shifts segregation in the caudal direction" (Toivonen and Saxén, 1968, p. 540).

This notion was obviously influenced by the distinction of Nieuwkoop et al. (1952) between an "activating" and a "transforming" principle, which in turn goes back to Needham and Waddington's assertion that in neural induction an initial step of "evocation," that is, generalized, nonspecific neural induction, is followed by a second step, "individualization," whereby regional specificity is determined (see above, Chapter 9, Neural and Mesodermal Inductions by Abnormal Inductors). My objection to this view is based on the fact that it is not supported by convincing experimental evidence; it may not even be possible, in prin-

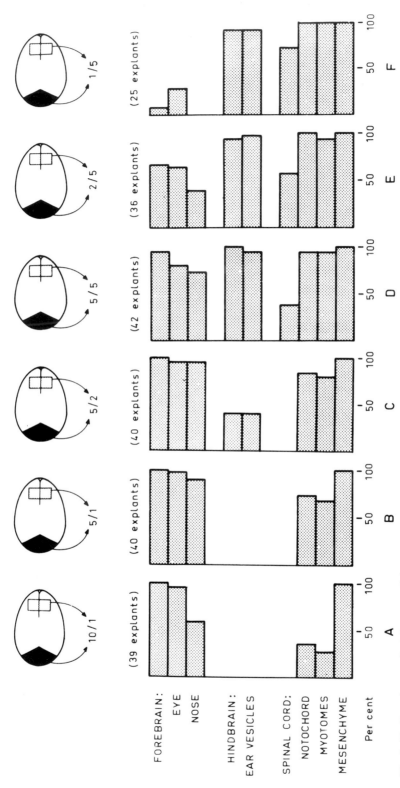

Fig. 9-11. Experiment of mixing neuralizing and mesodermizing embryonic tissues. Above, Archencephalic region of the neural plate *(black)* and the posterior mesoderm *(stippled)* of the neurula of *Triturus taeniatus*, and the ratios of the mixtures used in the dissociation and reaggregation experiments. Below, Frequency (in percent) of the different structures induced by the graded series of mixtures of the archencephalic and mesodermal inductors. From Toivonen and Saxén. 1968.

ciple, to prove this assertion. This would require the separation of the two steps by some experimental design. To my knowledge, no experiment on the induction of the nervous sytem has produced a clear case in which a generalized, not region-specific, but otherwise well-developed nervous system was differentiated. It is true that nonspecialized neural tubes were described in neural induction experiments. In most instances they were found in embryos that had been sacrificed too early, or they were the product of an ineffective inductor.

But there is another, more promising way of interpreting the regional specificity in terms of a two-step process. In pursuing it I follow another lead provided by Toivonen and Saxén. They asked very appropriately what the notion of a combined action of a neuralizing (N) agent and a mesodermizing (M) agent means in terms of the response of the reacting system, the ectoderm (Saxén and Toivonen, 1961, 1962). On the basis of experiments which I shall not discuss in detail they had come to the conclusion that when an N-agent and an M-agent reach the same area simultaneously, individual ectodermal cells or cell patches do not accept signals from both agents, but only from one; they are either neuralized or mesodermized. The two types of patches then interact with each other, and the outcome, in terms of regional specification, depends on the relative size of the patches. In the competition, the larger cell groups or larger numbers of approximately equal cell groups will prevail. "Due to the obvious mosaic-like distribution of the neuralizing and mesodermizing units on the contact surface, the overlying ectoderm will consequently be induced into intermixed islands of neural and mesodermal cells. The fate of these areas is then determined by subsequent conditions ... there are experimental data which demonstrate different inductive interactions during the period following primary induction" (Saxén and Toivonen, 1961, 528–529). In this sense, a two-step process of neural plate induction is very plausible. The significance of this new approach is obvious: one can now analyze neural plate induction in terms of cellular events within the framework of the double-gradient theory. The prospect that the biochemical identification of the N- and M-agents can be achieved and that their effects on the reacting system can be understood on the cellular and eventually on the molecular level no longer seems to be utopian.

We have come a long way from Spemann's discovery of the head and trunk organizer in 1931 to an understanding of organizer action in terms of the gradient distribution of two inductive agents three decades later. My intent in focusing on this particular legacy of Spemann has been stated repeatedly: to emphasize emphatically that embryonic induction is by no means the result of a nonspecific stimulus; rather, that it involves a demonstrable ingredient of "instruction," and that this major theme of classical experimental embryology remains a challenge to contemporary developmental biologists.

Finally, I hold out hope that this last chapter can clarify misunderstandings of the kind expressed in the following quotations: "I regard the misuse of concepts of induction as a major feature preventing progress in understanding pattern formation" (Wolpert, 1971, p. 184), and "Induction and its related concepts, which have so dominated embryological thinking, have completely obscured the problems of pattern formation by emphasizing the information coming from some other tissue rather than the response in the tissue which gives rise to the pattern." After a brief reference to the Spemann-Schotté experiment (Fig. 2–17), the quotation continues: "This illustrates the failure of inductive theory to consider the problem of spatial organization" (Wolpert, 1970, pp. 202–203).

If, instead of selecting the Spemann-Schotté experiment, the author had chosen as an example the induction of the neural plate—which by then had been proclaimed as *the* paradigm of embryonic induction—a different message would have emerged. It would have become evident that region-specific inductors can indeed create pattern. According to the double-gradient theory, the different regions of the neural plate along the main axis receive their positional information from the actions and interactions of two instruction-carrying inductive agents, the N-agent and the M-agent, which are distributed in the subjacent mesoderm in the form of two opposing gradients. To put it in more general terms: effect-specific induction, far from preventing an understanding of pattern formation, becomes in fact the creator of patterns by merging with another basic concept, the double-gradient concept. It is a major accomplishment of the double-gradient theory to provide a plausible, experimentally supported explanation of regional patterning of the neural plate by the juxtaposition of region-specific induction and gradients. Even if this linkage should be limited to the neural plate—which I believe it is not—it is obvious that the opinion expressed in the quotations cited above cannot be generalized. Induction and pattern formation can indeed live peacefully together!

Hilde Mangold,
Co-Discoverer of the Organizer

The now-famous paper by Spemann and Hilde Mangold, "On the Induction of Embryonic Anlagen by Implantation of Organizers from Different Species" (Ueber die Induktion von Embryonalanlagen durch Implantation artfremder Organisatoren), appeared in 1924 in W. Roux's *Archiv für Entwicklungsmechanik der Organismen,* at the time the most prestigious journal in the field of experimental embryology. Spemann, already a recognized leader in the field, achieved worldwide fame when in 1935 he received the Nobel Prize for this discovery. But who was Hilde Mangold? Very few of her contemporaries are still alive. As one of them who knew her well, I feel I should rescue her from oblivion. Though her name has disappeared from the literature, she deserves at least a footnote in the annals of experimental embryology.

She was an unusually gifted, vivacious, and charming young woman. Her considerable scientific talents would undoubtedly have borne fruit, had her life not been cut short by a tragic accident. She died of severe burns when a gasoline heater in her kitchen exploded. This occurred in September 1924, about the time when the organizer paper appeared in print. She was then twenty-six years old and the mother of an infant son, who was fated to die in his early twenties, a victim of Hitler's war. In the spring of 1924, the Mangolds had moved from Freiburg to Berlin-Dahlem, where Hilde's husband, Otto Mangold, the eldest student of Spemann, had been appointed head of the division of experimental embryology in the Kaiser Wilhelm (now Max Planck) Institute for Biology. In 1929, in the Festschrift for Spemann, Mangold published under his wife's name, posthumously, a brief report of her experiments in the spring of 1923, in which she had extended the organizer experiment to combinations of salamander species other than the ones used in the original experiment. Nothing more was ever heard of her.

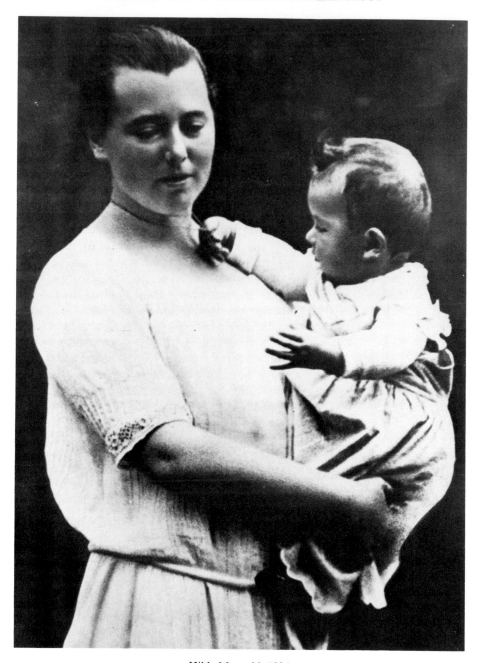

Hilde Mangold, 1924

What was the nature of the partnership of the two authors of the organizer paper? Quite simply, the experiment was the basis for the doctoral dissertation of Hilde Mangold, whose maiden name was Proescholdt; it was one of very few dissertations in the history of biology that was directly connected with a Nobel Prize. The conception and design of the experiment were Spemann's, based on a combination of ideas and experimental techniques that date back to his classic publication of 1918. Mangold's contribution was the execution of this difficult experiment, during the spring of 1921 and 1922.

Spemann had assumed the chairmanship of the Zoological Institute in Freiburg in the spring of 1919. He came from the Kaiser Wilhelm Institute in Berlin-Dahlem, where he had been codirector (with the botanist C. Correns, one of the rediscoverers of the Mendelian laws) and head of the division for experimental embryology from 1914 to 1918. Hilde Proescholdt and I arrived in Freiburg a year later, in the spring of 1920. She came from the University of Frankfurt, where she had attended a lecture of Spemann's that had intrigued her. I came from the University of Heidelberg, and my motives in moving to Freiburg were mixed. I had become interested in experimental embryology through C. Herbst, director of the Zoological Institute in Heidelberg, whose seminars I had attended. He and my aunt, Clara Hamburger (one of the first women to earn a Ph.D. in biology), a senior assistant at the Zoological Institute, assured me that Spemann had an excellent reputation. I was also attracted by the Black Forest mountains, which ascend to respectable heights immediately behind the city and offer much better opportunities for hiking and skiing than the lowly Odenwald near Heidelberg. In the spring of 1920 a friend and I had taken a hiking trip to Freiburg and the Feldberg, the highest mountain in the Black Forest; I had immediately felt an "elective affinity" for the area, which has not diminished to this day.

Soon after our arrival, a friendship developed between Hilde and me that was based on common interests in science, art, and literature. By nature she was open, frank, and cheerful. She had a penetrating and reflective intellect and a lively sense of beauty in nature and in the arts. Like myself she had grown up in a small town, loved the outdoors, and was familiar with plant and animal life. But she was perhaps at her best in the endless discussions and debates with kindred minds that extended through long evenings in the open-air taverns at the square around the cathedral or in our small rooms. When we were in danger of taking ourselves too seriously, it was her sense of humor that saved the situation. We were fortunate in that we had more time for such exercises than students today, because we did not have to prepare for examinations: there were neither midterms nor finals. We ourselves were responsible for what we learned or missed in a course. But then, we were at the level of seniors or graduate students and had received most of our college-level education at the gymnasium.

We cared more about food for thought than about nourishment for our bodies. These were the early, turbulent years of the Weimar Republic—the aftermath of the war and the Versailles treaty, which fomented political strife and unrest in Germany. As usual, the extremists clashed violently at the universities. Most of the fraternities, which played an important role in university life, were allied with the extreme right. Their superpatriotic, often vicious attacks were directed as much against the democratic republic as against their communist counterparts on the extreme left. The liberal democrats in the center, with whom Hilde and I sympathized, had a hard time defending the republic.

Politics was not our major concern, however. What had brought us to Freiburg was the quest of science. Soon after our arrival Hilde Proescholdt and I presented ourselves to Spemann, and we were accepted as candidates for the Ph.D. In those golden days no special examinations or other formalities were required; probably we looked mature and eager, and we had recommendations from our previous teachers. We were assigned adjacent workbenches in a large laboratory room on the second floor of the modest Zoological Institute. This institute dated back to the 1880s. Its founder and first director was August Weismann, one of the leading Darwinists of his epoch (his book on the "germ plasm" had inspired Spemann's first experiments on salamander eggs). The first floor was occupied by the offices and laboratories of Spemann and his two assistants, F. Baltzer and O. Mangold. Hilde and I were joined later by Hans Holtfreter, and there were two or three other doctoral candidates. Holtfreter later became one of my best friends and, more important, the leading experimental embryologist of his generation. At that time we had little personal contact.

Holtfreter was an ardent member of the youth movement, which had its stronghold in the universities. Like youth movements at other times and in other countries, it expressed in its life-style its opposition to the bourgeoisie. The German version was a romantic "back to nature" movement that revived old Germanic folklore, folksongs, and dances. It became almost a cult. Solemn regional and national assemblies were held, at which the huge evening campfire was the ceremonial center. Both sexes joined in, and unlike our Boy Scouts and Girl Scouts, parents and the older generation were excluded altogether. The young people called themselves "Wandervögel" (migrant birds), and indeed they spent all their free time hiking with knapsacks to the mountains and seashores of the fatherland and to foreign countries.

Hilde Proescholdt and I were less adventurous. We were more attracted to the intellectual and artistic movements flourishing in the young Weimar Republic. We were attuned to the poetry of Rainer Maria Rilke and Stefan George and to German expressionist art, which challenged the traditional aesthetic values and tried to abrogate the laws of nature. The walls of our rooms were decorated with the "Blue Horses" of Franz Marc, as well as with the graphic art of the German masters Dürer and Cranach. We sensed the affinity of expressionism and the medieval art that surrounded us in Freiburg. Its cathedral (the Muenster) is one of the most beautiful medieval masterpieces in Germany, richly decorated with sculptures, paintings, and grotesque gargoyles. The center of the town, with the grain merchant's house and the archbishop's residence at opposite sides of the cathedral square, the narrow winding streets, and the two city gates, preserved a distinctly medieval touch. And the atmosphere of the open-air market around the Muenster on Saturdays, when the women of the countryside set up their stands under colorful umbrellas, offering their home-grown fruit and vegetables and homemade bread and cheeses probably had scarcely changed in centuries. Some of the women still wore their native garb, distinctive for each Black Forest valley, with different embroideries on their waistcoats and different shapes of straw hats with long black ribbons. The annual Fronleichnam (Corpus Christi) procession in June again continued an age-old tradition. The old guilds marched behind their banners; the university professors (there was a Catholic theological faculty) and even the otherwise rather worldly Verbindungen (fraternities) in their paramilitary uniforms joined the archbishop and priests and the many faithful. The procession wound its way through the town, stopping at several richly decorated open-air altars, where mass was read to the sound of church bells all over the town.

It was one of the blessings of student life in those days that we had an unlimited choice of courses. As predoctoral students we followed no curriculum, but elected courses according to our interests.

One of our favorite courses was that of Professor Grosse, a sensitive interpreter of Japanese and Chinese art. A former resident of Japan, he was married to a Japanese, and in his deportment presented the quiet dignity of the Japanese elders. I have not forgotten the occasion when he took from his pocket a small object wrapped in a precious silk cloth; he unwound it with tender care and held it up for us to appreciate. It was a netsuke, a miniature artwork carved in wood or ivory, which the Japanese used to carry in the sash of their kimono like a jewel. His course was attended not only by students of art history and the humanities, but also by many scientists and even medical and premedical students— who at that time could afford to cultivate such interests.

We were introduced to philosophy by one of its most eminent representatives, Professor E. Husserl, the founder of phenomenology and the teacher of M. Heidegger, who became his successor in 1929. (I was back in Freiburg that year and attended one of Heidegger's seminars; I befriended his assistant, W. Brock, who held a seminar with a strong Aristotelian flavor on the philosophy of Biology.) Philosophy then still held its traditional central rank at the university, and those of us who earned the degree of Doctor of Philosophy knew at least what philosophy was about and had respect for it. Times have changed!

The intense intellectual and spiritual life was perhaps our salvation from the miserable living conditions of the postwar years, aggravated by a staggering inflation. Everyone who survived this period remembers the "turnip winter," when we practically existed on this staple alone—and never touched it again. Another memory is that of the Saturday extravaganza, to which we looked forward, when the student cafeteria (then more aptly called the "mensa") served us hot chocolate with milk and an unlimited number of buns, donated by the American Friends Service Committee.

Hilde Proescholdt and I shared a vivid interest in the rich flora of the environs of Freiburg. The high mountains of the Black Forest harbored alpine plants and glacial relics, gentians, alpine primroses, and saxifrages. Toward the west, an unusual Mediterranean plant community had settled in the warm and sunny Rhine Valley. In early spring the slopes were covered with white and blue anemones. Our most precious finds were rare orchids in the woods and hillsides of the Kaiserstuhl, a volcanic intrusion that rises like an island out of the Rhine Valley. The vineyards there produce the best local wines. One of their connoisseurs, our botany professor, F. Oltmanns (an elderly bachelor and bon vivant), familiarized us not only with the Latin names of many rare plant species and their long journeys from remote regions of Europe to this idyllic spot, but also with his favorite inns in the small towns and villages and their special homegrown vintages.

Hilde Proescholdt, like me, came from a fairly well-to-do merchant family. Her hometown of Gotha, in Thuringia, had been the seat of the Grand Duke of Saxony and Gotha; he had been in residence until a few years previously and there was, of course, a castle overlooking the town. Unfortunately, no biographical material about Hilde is available, but I found two of her letters, both from Gotha, dated August and December 1920, which miraculously survived the dislocations of six decades. She was an ardent reader, and her interests had a remarkable range. On her Christmas vacation she wrote: "Scheler, Husserl, Gundolf, George, and two Worringer books give me a lot to do." (Scheler and Husserl were the leading philosophers of phenomenology; Gundolf was a prominent member of the esoteric circle around the poet Stefan George; and Worringer was a widely discussed and influential art historian.) "Then an interesting book by Volkelt: 'The Aesthetic Consciousness.' Found it quite by accident in the library of our castle. Up there it is fine. There, in a very old tower with thick walls, sit all the bookworms. ... Now Volkelt: in contrast to Husserl he spins threads of connections between phenomenology and psychology. He turns up very fine trends of thought." Then follows a discourse on empathy in art, on the part of both artist and viewer, with critical remarks on the viewpoints of Volkelt and Worringer. She then turns the discussion to romanesque art of the time of Charlemagne, preserved in old village churches on an island in Lake Constance that we had visited together. Included in the letter were newspaper clippings—portraits of a Roman man and woman— which had elicited reflections on the relations of romanesque art and the art of Rome and speculations that the monks of the time, in South Germany, may have had direct connections with Rome.

In the other letter she told of reading Th. Boveri. (He and the American E. B. Wilson were the leading cytologists and chromosome specialists who originated the chromosome theory of heredity. Both Spemann and Baltzer were students of Boveri.) She wrote: "The lecture course of Baltzer [on genetics] was rather bad. I am sorry that I have to write that, but I cannot say otherwise, after objective comparison with the original works. He did not give any plasticity. I wonder whether you understand. The essentials were not emphasized and never sharply illuminated. Should not a Dozent do that? The pleasure despite all this was in the confrontation with these problems." She was probably right.

Baltzer redeemed himself in the following semester, when he consented to give Hilde Proescholdt and me a special laboratory course in cytology. The emphasis was on chromosomes and their behavior in cell division and gamete formation; insects were the material of choice. This was a do-it-yourself adventure. We caught grasshoppers and bugs in the field; we dissected the gonads and prepared beautifully stained microscope slides, following religiously the recipes in the laboratory manual of P. Buchner, who was not only an expert but an artist who did the beautiful illustrations in his book. We also had access to Baltzer's slides of the chromosomes of the roundworm, Ascaris. This lowly creature had risen to fame, thanks to the classic studies of Baltzer's teacher, Boveri, who had taken advantage of the small number of its chromosomes—just four. (Drosophila, now the indispensable household staple for cytogenetic studies, had been "discovered" by T. H. Morgan only a few years earlier and was not yet available.)

Lecture courses took only a small part of our time. Most of our waking hours were spent at our workbenches and devoted to our experiments. Long before the breeding season of the amphibians began early in April, we were at work, preparing the delicate glass instruments that Spemann had invented for doing microsurgery on the tiny embryos. The operations were performed with glass needles, drawn out to an almost microscopic point in the flame of a microgasburner that had been fashioned by extending a tube to capillary size. With these glass needles we cut out tiny fragments of the soft embryo. Steady hands were required, because we worked under a binocular low-power microscope that also magnified our movements. The fragment was then transferred to another embryo in a micropipette and implanted at a specified site. The embryos were moved and turned over with a "hairloop," another clever invention of Spemann's: a soft baby's hair (originally taken from his own son) was looped and the two ends were fitted into the capillary opening of a handle that consisted of a piece of glass tubing, then sealed in with hot wax.

After some practice, doing the experiments was fun. But we had to cope with a very high mortality; the embryos, once removed from their protective jelly membranes, were extremely delicate, and at that time we had no weapon against bacterial infection. It took most of us two years to complete an experiment. Proescholdt had a particularly difficult job, because her experiment involved transplantation from embryos of one species to those of another species, thereby increasing the risk. Her plight, and also her stamina and perseverance, are attested by the fact that of her many experiments over two breeding seasons, only six embryos were eventually found worthy of being presented to the general public in the famous organizer paper.

There were vacations, too. In March 1921, after the end of the winter semester, Baltzer, a native Swiss and an expert skier, became our ski-master. We "Doctoranden" and the younger instructors spent a couple of weeks with him in the Black Forest, reveling in the snow and sunshine of the late winter. Hilde Proescholdt and Otto Mangold were in our group. They were married the same year and in the spring of 1924 moved to Berlin-Dahlem.

The story of Hilde Proescholdt's dissertation would not be complete without recounting some whimsical twists. While practically all of Spemann's students were given problems dealing with the early development of amphibians, Spemann made an exception with Hilde. His thoughts had turned back to one of his famous predecessors in experimental zoology, the French amateur naturalist Abraham Trembley. In the late eighteenth century Trembley had discovered the amazing power of regeneration of the freshwater polyp, Hydra. He had cut the animal into small pieces, and each of them had regenerated a complete polyp. Among the many ingenious experiments he had performed was one that had particularly intrigued Spemann. He had managed to turn the polyp inside out, contending that eventually the outer lining of the body was transformed into the lining of the intestinal tube and vice versa. The interchangeability of the two layers had a parallel: the transplantation experiments of Otto Mangold had shown the same interchangeability of inner and outer germ layers in early amphibian embryos. Apparently Spemann had then become interested in checking on Trembley's claim, and he had suggested to Hilde Proescholdt that she repeat the inversion experiment. She set out to do so. Despite her considerable skill and perseverance, she was not successful. Even the help of the master of microsurgery himself was of no avail; the two of them tried to hold the inverted hydra in place with a fine glass rod, but the uncooperative creature always managed to uncurl. In the meantime, the relatively short breeding season of the amphibians had advanced (methods for inducing egglaying by hormone treatment were not yet available), and Proescholdt became impatient. Spemann accommodated her and turned over to her an experiment that had a high priority on his agenda: the transplantation of the upper lip of the blastopore of an early gastrula to the flank of a gastrula of another species. With beginner's luck, she obtained in early May an embryo which displayed on its flank a large secondary neural tube. Spemann—and everybody else in the laboratory—was impressed, and this one case was reported in a brief postscript dated May 1921 to a Spemann publication of the same year that dealt with transplantations between species (obviously a related topic). The term "organizer" was introduced and defined in this postscript. But it took Proescholdt another breeding season to complete her thesis work. The results were written up jointly by the two authors; the paper was submitted to Roux's *Archiv* in June 1923 and it appeared in print in 1924.

Hilde Proescholdt, who in the meantime had become Mrs. Mangold, was not happy that Spemann had added his name to her thesis publication, while Holtfreter and I and all the rest of us saw ourselves proudly in print as sole authors. Moreover, Spemann had insisted on having his name precede hers! But Spemann was perfectly right in claiming precedence, while she apparently did not fully realize the significance of her results. It was not granted to her to live to see the great impact her experiment had on the course of experimental embryology.

References

Adelmann, H. B. (1929). Experimental studies on the development of the eye. II. The eye-forming potencies of the median portions of the urodelan neural plate (*Triton taeniatus* and *Amblystoma punctatum*). *J. Exp. Zool.*, *54*, 291–317.

Adelmann, H. B. (1930). Experimental studies on the development of the eye. III. The effect of the substrate ("Unterlagerung") on the heterotopic development of the median and lateral strips of the anterior end of the neural plate of *Amblystoma*. *J. Exp. Zool.*, *57*, 223–281.

Adelmann, H. B. (1932). The development of the prechordal plate and mesoderm of *Amblystoma punctatum*. *J. Morphol.*, *54*, 1–67.

Adelmann, H. B. (1936). The problem of cyclopia. *Quart. Rev. Biol.*, *11*, 161–182, 284–304.

Alderman, A. L. (1935). The determination of the eye in the anuran, *Hyla regina*. *J. Exp. Zool.*, *70*, 205–232.

von Baer, C. E. (1828). *Ueber Entwicklungsgeschichte der Thiere. Beobachtung und Reflexion.* Gebrüder Bornträger, Königsberg. Reprinted 1967, Culture et Civilisation, Brussels.

Balinsky, B. I. (1978). *An Introduction to Embryology,* Fourth Edition. W. B. Saunders Company, Philadelphia.

Baltzer, F. (1952). Experimentelle Beiträge zur Homologie. Xenoplastische Transplantationen bei Amphibien. *Experientia, 8,* 285–297.

Baltzer, F. (1962). *Theodor Boveri.* Wissenschaftliche Verlagsgesellschaft, M.B.H., Stuttgart. English translation, 1967, University of California Press, Berkeley and Los Angeles.

Barth, L. G. (1941). Neural differentiation without organizer. *J. Exp. Zool., 87,* 371–384.

Bautzmann, H. (1926). Experimentelle Untersuchungen zur Abgrenzung des Organizationszentrums bei *Triton taeniatus.* *Roux' Arch. f. Entw. mech., 108,* 283–321.

Bautzmann, H. (1929). Über bedeutungsfremde Selbstdifferenzierung aus Teilstücken des Amphibienkeimes. *Naturwissenschaften, 17,* 818–827.

Bautzmann, H., Holtfreter, J., Spemann, H. and Mangold, O. (1932). Versuche zur Analyse der Induktionsmittel in der Embryonalentwicklung. *Naturwissenschaften, 20,* 971–974.

Bijtel, H. (1931). Über die Entwicklung des Schwanzes bei Amphibien. *Roux' Arch f. Entw. mech., 125,* 448–486.

Bruns, E. (1931). Experimente über das Regulationsvermögen der Blastula von *Triton taeniatus* und *Bombinator pachypus. Roux' Arch. f. Entw. mech., 123,* 682–718.

Child, C. M. (1929). Physiological dominance and physiological isolation in development and reconstitution. *Roux' Arch. f. Entw. mech., 117,* 21–66.

Child, C. M. (1941). *Patterns and Problems of Development*. University of Chicago Press, Chicago.

Child, C. M. (1946). Organizers in development and the organizer concept. *Physiol. Zool., 19,* 89–148.

Chuang, H.-H. (1938). Spezifische Induktionsleistungen von Leber und Niere im Explantatversuch. *Biol. Zentralbl., 58,* 472–480.

Chuang, H.-H. (1939). Induktionsleistungen von frischen und gekochten Organteilen (Niere, Leber) nach ihrer Verpflanzung in Explantate und verschiedene Wirtsregionen von Tritonkeimen. *Roux' Arch. f. Entw. mech., 139,* 556–638.

Chuang, H.-H. (1940). Weitere Versuche über die Veränderung der Induktionsleistungen von gekochten Organteilen. *Roux' Arch. f. Entw. mech., 140,* 25–38.

Copenhaver, W. M. (1926). Experiments on the development of the heart of *Amblystoma punctatum. J. Exp. Zool., 43,* 321–371.

Curtis, A. S. G. (1960). Cortical grafting in *Xenopus laevis. J. Embr. exp. Morphol., 8,* 163–173.

Driesch, H. (1891). Entwicklungsmechanische Studien I. Der Wert der beiden ersten Furchungszellen in der Echinodermenentwicklung. Experimentelle Erzeugung von Teil- und Doppelbildungen. *Zeitschr. wiss. Zool., 53,* 160–183.

Driesch, H. (1894). *Analytische Theorie der Organischen Entwicklung.* W. Engelmann, Leipzig.

Driesch, H. (1951). *Lebenserinnerungen.* Ernst Reinhardt Verlag, München/Basel.

Dürken, B. (1926). Das Verhalten embryonaler Zellen im Implantat. *Roux' Arch. f. Entw. mech., 107,* 727–828.

Ekman, G. (1925). Experimentelle Beiträge zur Herzentwicklung der Amphibien. *Roux' Arch. f. Entw. mech., 106,* 320–352.

Endres, H. (1895). Über Anstich- und Schnürversuche an Eiern von *Triton taeniatus. Schles. Gesellsch. vaterländ. Kultur., 73.*

Filatow, D. (1925). Über die unabhängige Entstehung (Selbstdifferenzierung) der Linse bei *Rana esculenta. Roux' Arch. f. Entw. mech., 104,* 50–71.

Fischer, F. G. und Wehmeier, E. (1933a). Zur Kenntnis der Induktionsmittel in der Embryonalentwicklung. *Naturwissenschaften, 21,* 518.

Fischer, F. G. und Wehmeier, E. (1933b). Zur Kenntnis der Induktionsmittel in der Embryonalentwicklung. *Nachr. Ges. d. Wiss. Göttingen, VI. Biol. 9,* 394–400.

Geinitz, B. (1925). Embryonale Transplantation zwischen Urodelen und Anuren. *Roux' Arch. f. Entw. mech., 106,* 357–408.

Gerhart, J. C. (1980). Mechanisms regulating pattern formation in the amphibian egg and early embryo. In: *Biological Regulation and Development,* Vol. 2, ed. R. F. Goldberger, pp. 133–316. Plenum Press, New York.

Gilbert, S. F. (1985). *Developmental Biology.* Sinauer Associates, Sunderland, MA.

Gimlich, R. L. (1985). Cytoplasmic localization, inductions, and the organizer of the frog embryo. In: *Molecular Determinants of Animal Form.* Proc. U.C.L.A. Symposia, pp. 15–36.

Goerttler, K. (1925). Die Formbildung der Medullarplatte bei Urodelen, im Rahmen der Verschiebungsvorgänge von Keimbezirken während der Gastrulation. *Roux' Arch. f. Entw. mech., 106,* 503–541.

Goerttler, K. (1926). Experimentell erzeugte *"Spina bifida"* und Ringembryonenbildungen und ihre Bedeutung für die Entwicklungsphysiologie der Urodeleneier. *Zeitschr. f. Anat. Entw., 80,* 283–343.

Goerttler, K. (1927). Die Bedeutung gestaltender Bewegungsvorgänge beim Differenzierungsgeschehen. *Roux' Arch. f. Entw. mech., 112,* 518–576.

Gurwitsch, A. (1922). Über den Begriff des embryonalen Feldes. *Roux' Arch. f. Entw. mech., 51,* 383–415.

Hadorn. E. (1965). Problems of determination and transdetermination. *Brookhaven Symp. Biol., 18,* 148–161.

Haeckel, E. (1891). *Anthropogenie oder Entwicklungsgeschichte des Menschen,* 4th rev. and enl. ed. Wilhelm Engelmann, Leipzig.

Hamburger, V. (1926). Versuche über Komplementärfarben bei Ellritzen *(Phoxinus laevis). Zeitschr. f. vergl. Physiol., 4,* 286–304.

Hamburger, V. (1960). *A Manual of Experimental Embryology*, revised ed. University of Chicago Press, Chicago and London.

Hamburger, V. (1969). Hans Spemann and the organizer concept. *Experientia, 25*, 1121–1125.

Hamburger, V. (1980a). Embryology and the modern synthesis in evolutionary theory. In: *The Evolutionary Synthesis*, ed. E. Mayr and W. Provine, pp. 96–112. Harvard University Press, Cambridge.

Hamburger, V. (1980b). S. Ramon y Cajal, R. G. Harrison and the beginnings of neuroembryology. *Perspect. Biol. Med., 23*, 601–616.

Hamburger, V. (1985). Hans Spemann, Nobel Laureate 1935. *Trends Neurosci., 8*, 385–387.

Harrison, R. G. (1903). Experimentelle Untersuchungen über die Entwicklung der Sinnesorgane der Seitenlinie bei den Amphibien. *Arch. f. mikr. Anat., 63*, 35–149.

Harrison, R. G. (1918). Experiments on the development of the forelimb of *Amblystoma*, a self-differentiating, equipotential system. *J. Exp. Zool., 25*, 413–461.

Harrison, R. G. (1921). On relations of symmetry in transplanted limbs. *J. Exp. Zool., 32*, 1–136.

Harrison, R. G. (1933). Some difficulties of the determination problem. *Am. Naturalist, 67*, 306–321.

Harrison, R. G. (1935). Heteroplastic grafting in embryology. *Harvey Lectures 1933–34*, 116–157.

Harrison, R. G. (1937). Embryology and its relations. *Science, 85*, 369–374.

Harrison, R. G. (1945). Relations of symmetry in the developing embryo. *Trans. Conn. Acad. Arts Sci., 36*, 277–330.

Harrison, R. G. (1969). *Organization and Development of the Embryo*, ed. S. Wilens. Yale University Press, New Haven and London.

Herbst, C. (1893). Experimentelle Untersuchungen über den Einfluss der veränderten chemischen Zusammensetzung des umgebenden Mediums auf die Entwicklung der Tiere. II. Weiteres über die Wirkung der Lithiumsalze. *Mitt. Zool. Stat. Neapel., 11*, 136–220.

Herbst, C. (1901). *Formative Reize in der tierischen Ontogenese*. Arthur Georgi, Leipzig.

Herlitzka, A. (1897). Sullo sviluppo di embrioni completi da blastomeri isolati di uova di Tritone. *Roux' Arch. f. Entw. mech., 4*, 624–658.

Hertwig, O. (1893). Über den Wert der ersten Furchungszellen für die Organbildung des Embryo. *Arch f. Mikr. Anat., 42*, 662–806.

His, W. (1874). *Unsere Köperform und das physiologische Problem ihrer Entstehung*. R. C. W. Vogel, Leipzig.

Holtfreter, H. B. (1965). Differentiation capacities of Spemann's organizer investigated in explants of diminishing size. Ph.D. Thesis, University of Rochester, 1–252.

Holtfreter, J. (1925). Defekt- und Transplantationsversuche an der Anlage von Leber und Pancreas jüngster Amphibienkeime. *Roux' Arch. f. Entw. mech., 105*, 330–384.

Holtfreter, J. (1929a). Über histologische Differenzierungen von isoliertem Material jüngster Amphibienkeime. *Verh. d. deutschen Zool. Ges.*, 174–181.

Holtfreter, J. (1929b). Über die Aufzucht isolierter Teile des Amphibienkeimes. I. *Roux' Arch. f. Entw. mech., 117*, 421–510.

Holtfreter, J. (1931a). Über die Aufzucht isolierter Teile des Amphibienkeimes. II. *Roux' Arch. f. Entw. mech., 124*, 404–466.

Holtfreter, J. (1931b). Potenzprüfungen am Amphibienkeim mit Hilfe der Isolationsmethode. *Verh. d. deutschen Zool. Ges.*, 158–166.

Holtfreter, J. (1933a). Nicht typische Gestaltungsbewegungen, sondern Induktionsvorgänge bedingen medullare Entwicklung von Gastrulaektoderm. *Roux' Arch. f. Entw. mech., 127*, 591–618.

Holtfreter, J. (1933b). Der Einfluss von Wirtsalter und verschiedenen Organbezirken auf die Differenzierung von angelagertem Gastrulaektoderm. *Roux' Arch. f. Entw. mech., 127*, 610–775.

Holtfreter, J. (1933c). Organisierungsstufen nach regionaler Kombination von Entomesoderm mit Ektoderm. *Biol. Zentralbl., 53*, 404–431.

Holtfreter, J. (1933d). Eigenschaften und Verbreitung induzierender Stoffe. *Naturwissenschaften, 21*, 766–770.

Holtfreter, J. (1933e). Nachweis der Induktionsfähigkeit abgetöteter Keimteile. *Roux' Arch. f. Entw. mech., 128,* 584–633.

Holtfreter, J. (1933f). Die totale Exogastrulation, eine Selbstablösung des Ektoderms vom Entomesoderm. *Roux' Arch. f. Entw. mech., 129,* 669–793.

Holtfreter, J. (1934a). Formative Reize in der Embryonalentwicklung der Amphibien, dargestellt an Explantationsversuchen. *Arch. exp. Zellf., 15,* 281–301.

Holtfreter, J. (1934b). Der Einfluss thermischer, mechanischer und chemischer Eingriffe auf die Induzierfähigkeit von Triton-Keimteilen. *Roux' Arch. f. Entw. mech., 132,* 225–306.

Holtfreter, J. (1934c). Über die Verbreitung induzierender Substanzen und ihre Leistungen im Triton-Keim. *Roux' Arch. f. Entw. mech., 132,* 307–383.

Holtfreter, J. (1936). Regionale Induktionen in xenoplastisch zusammengesetzten Explantaten. *Roux' Arch. f. Entw. mech., 134,* 446–550.

Holtfreter, J. (1938a). Differenzierungspotenzen isolierter Teile der Urodelengastrula. *Roux' Arch. f. Entw. mech., 138,* 522–656.

Holtfreter, J. (1938b). Differenzierungspotenzen isolierter Teile der Anurengastrula. *Roux' Arch. f. Entw. mech., 138,* 657–738.

Holtfreter, J. (1939). Gewebeaffinität, ein Mittel der embryonalen Formbildung. *Arch. exp. Zellf., 23,* 169–209.

Holtfreter, J. (1944). Neural differentiation of ectoderm through exposure to saline solution. *J. Exp. Zool., 95,* 307–340.

Holtfreter, J. (1945). Neuralization and epidermization of gastrula ectoderm. *J. Exp. Zool., 98,* 161–209.

Holtfreter, J. (1951). Orientaciones modernas de la Embryologia. *Inst. de Invest. Cienc. Biol. (Montevideo, Uruguay), 1,* 285–349.

Holtfreter, J. (1982). Recollections of an embryologist—in search of the organizer. *Shizen* [*Nature*], *82,* 48–61 (in Japanese).

Holtfreter, J. and Hamburger, V. (1955). Amphibians. In: *Analysis of Development,* ed. B. H. Willier, P. Weiss, and V. Hamburger, pp. 230–296. W. B. Saunders Company, Philadelphia and London.

Horder, T. J. and Weindling, P. J. (1986). Hans Spemann and the organiser. In: *A History of Embryology,* ed. T. J. Horder, J. A. Witkowski, and C. C. Wylie, pp. 183–242. Cambridge University Press, Cambridge.

Hörstadius, S. (1939). The mechanics of sea urchin development, studied by operative methods. *Biol. Rev., 14,* 132–179.

Hörstadius, S. (1973). *Experimental Embryology of Echinoderms.* Clarendon Press, Oxford.

Huxley, J. S. and De Beer, G. R. (1934). *The Elements of Experimental Embryology.* Cambridge University Press, Cambridge.

Jacobson, A. G. (1966). Inductive processes in embryonic development. *Science, 152,* 25–34.

Kingsbury, B. F. and Adelmann, H. B. (1924). The morphological plan of the head. *Quart. J. Micr. Sci., 68,* 239–285.

Kirschner, M. W. and Gerhart, J. C. (1981). Spatial and temporal changes in the amphibian egg. *Bio-Science, 31,* 381–388.

Krämer, W. (1934). Über Regulations- und Induktionsleistungen destruierter Induktoren. *Roux' Arch. f. Entw. mech., 131,* 220–237.

Kusche, W. (1929). Interplantation umschriebener Zellbezirke aus der Blastula und Gastrula der Amphibien. I. Versuche an Urodelen. *Roux' Arch. f. Entw. mech., 120,* 192–271.

Lehmann, F. E. (1926). Entwicklungsstörungen an der Medullaranlage von *Triton,* erzeugt durch Unterlagerungsdefekte. *Roux' Arch. f. Entw. mech., 108,* 243–282.

Lehmann, F. E. (1928). Die Bedeutung der Unterlagerung für die Entwicklung der Medullarplatte von *Triton. Roux' Arch. f. Entw. mech., 113,* 123–171.

Lehmann, F. E. (1937). Die Wirkungsweise chemischer Faktoren in der Embryonalentwicklung der Tiere. *Rev. Suisse de Zool., 44,* 1–20.

Lehmann, F. E. (1942). Spezifische Stoffwirkungen bei der Induktion des Nervensystems der Amphibien. *Naturwissenschaften, 30,* 515–526.

Lehmann, F. E. (1945). *Einführung in die Physiologische Embryologie.* Birkhäuser Verlag, Basel.

Lehmann, F. E. (1950). Die Morphogenese in ihrer Abhängigkeit von elementaren biologischen Konstituenten des Plasmas. *Rev. Suisse de Zool., 57, Suppl. 1,* 141–151.

Lewis, W. H. (1904). Experimental studies on the development of the eye in amphibia. I. On the origin of the lens in *Rana palustris. Am. J. Anat., 3,* 505–536.

Lewis, W. H. (1907). Experimental studies on the development of the eye in amphibia. III. On the origin and differentiation of the lens. *Am. J. Anat., 6,* 473–509.

Mangold, H., geb. Pröscholdt. (1929). Organisatortransplantationen in verschiedenen Kombinationen bei Urodelen. *Roux' Arch. f. Entw. mech., 117,* 697–710.

Mangold, O. (1920). Fragen der Regulation und Determination an umgeordneten Furchungsstadien und verschmolzenen Keimen von *Triton. Roux' Arch. f. Entw. mech., 47,* 250–301.

Mangold, O. (1923). Transplantationsversuche zur Frage der Specifizität und der Bildung der Keimblätter bei *Triton. Roux' Arch. f. Entw. mech., 100,* 198–301.

Mangold, O. (1925). Die Bedeutung der Keimblätter in der Entwicklung. *Naturwissenschaften., 13,* 213–218.

Mangold, O. (1929). Experimente zur Analyse der Determination and Induktion der Medullarplatte. *Roux' Arch. f. Entw. mech., 117,* 586–696.

Mangold, O. (1931). Das Determinationsproblem. III. Das Wirbeltierauge in der Entwicklung und Regeneration. *Erg. d. Biol., 7,* 193–403.

Mangold, O. (1933). Über die Induktionsfähigkeit der verschiedenen Bezirke der Neurula von Urodelen. *Naturwissenschaften, 21,* 761–766.

Mangold, O. (1953). *Hans Spemann, ein Meister der Entwicklungsphysiologie.* Wissenschaftliche Verlagsgesellschaft M. B. H., Stuttgart.

Mangold, O. und Seidel, F. (1927). Homoplastische und heteroplastische Verschmelzung ganzer Tritonkeime. *Roux' Arch. f. Entw. mech., 111,* 593–665.

Mangold, O. und Spemann, H. (1927). Über Induktion von Medullarplatte durch Medullarplatte im jüngeren Keim. *Roux' Arch. f. Entw. mech., 111,* 341–422.

Mangold, O. und von Woellwarth, C. (1950). Des Gehirn von Triton. Ein experimenteller Beitrag zur Analyse seiner Determination. *Naturwissenschaften, 37,* 365–372.

Marx, A. (1925). Experimentelle Untersuchungen zur Frage der Determination der Medullarplatte. *Roux' Arch. f. Entw. mech., 105,* 20–44.

Mayer, B. (1935). Über das Regulations- and Induktionsvermögen der halbseitigen oberen Urmundlippe von *Triton. Roux' Arch. f. Entw. mech., 133,* 518–581.

Mayr, E. (1982). *The Growth of Biological Thought.* Harvard University Press, Cambridge.

Mencl, E. (1903). Ein Fall von beiderseitiger Augenlinsenausbildung während der Abwesenheit von Augenblasen. *Roux' Arch. f. Entw. mech., 16,* 328–339.

Morgan, T. H. (1895). Half embryos and whole embryos from one of the first two blastomeres. *Anat. Anz., 10,* 623–628.

Murray, P. D. F. (1928). Chorio-allantoic grafts of fragments of the two-day chick, with special reference to the development of the limbs, intestine and skin. *Austral. J. Exp. Biol. Med. Sci.,* 237–256.

Nakamura, O. (1978). Epigenetic formation of the organizer. In: *Organizer—a Milestone of Half-Century from Spemann,* ed. O. Nakamura and S. Toivonen, pp. 179–220. Elsevier/North-Holland Biomedical Press, Amsterdam.

Needham, J. (1942). *Biochemistry and Morphogenesis.* Cambridge University Press, Cambridge.

Needham, J. (1968). Organizer phenomena after four decades: a retrospect and prospect. In: *J. B. S. Haldane and Modern Biology,* ed. K. R. Dronamraju, pp. 227–298. Johns Hopkins Press, Baltimore.

Needham, J., Waddington, C. H. and Needham, D. M. (1934). Physico-chemical experiments on the amphibian organizer. *Proc. Royal Soc. (Lond.) B, 114,* 393–422.

Nieuwkoop. P. D. (1969). The formation of the mesoderm in urodelean amphibians. I. Induction by the endoderm. *Roux' Arch. f. Entw. mech., 162,* 341–373.

Nieuwkoop, P. D. (1973). The 'organization center' of the amphibian embryo, its origin, spatial organization and morphogenetic action. *Advances in Morphogenesis, 10,* 1–39.

Nieuwkoop, P. D. et al. (1952). Activation and organization of the central nervous system. I.

Induction and activation. II. Differentiation and organization. III. Synthesis of a new working hypothesis. *J. Exp. Zool.*, *120*, 1–108.

Nobel Foundation, ed. (1962). *Nobel, The Man and His Prizes*. Elsevier Publishing Company, Amsterdam.

Okada, T. (1960). Epithelio-mesenchymal relationships in the regional differentiation of the digestive tract in the amphibian embryo. *Roux' Arch. f. Entw. mech.*, *152*, 1–21.

Okada, T. (1980). Cellular metaplasia or transdifferentiation as a model for retinal cell differentiation. *Curr. Topics Dev. Biol.*, *16*, 349–380.

Oppenheimer, J. (1967). Analysis of development: problems, concepts and their history. In: *Essays in the History of Embryology and Biology*, pp. 117–172. M.I.T. Press, Cambridge.

Oppenheimer, J. (1970). Some diverse backgrounds for Curt Herbst's ideas about embryonic induction. *Bull. Hist. Med.*, *44*, 241–250.

Puck, T. T., Cieciura, S. J. and Robinson, A. (1958). Genetics of somatic mammalian cells. III. Long-term cultivation of euploid cells from human and animal subjects. *J. Exp. Med.*, *108*, 948–956.

Rhumbler, L. (1897). Stemmen die Strahlen der Astrosphaere oder ziehen sie? *Roux' Arch f. Entw. mech.*, *4*, 659–730.

Rotmann, E. (1931). Die Rolle des Ektoderms und Mesoderms bei der Formbildung der Kiemen und Extremitäten von *Triton*. *Roux' Arch. f. Entw. mech.*, *124*, 747–794.

Rotmann, E. (1933). Die Rolle des Ektoderms und Mesoderms bei der Formbildung der Extremitäten von *Triton*. II. Operationen im Gastrula- und Schwanzknospenstadium. *Roux' Arch. f. Entw.mech.*, *129*, 85–119.

Rotmann, E. (1942). Zur Frage der Leistungsspezifität abnormer Induktoren. *Naturwissenschaften*, *30*, 60–62.

Roux, W. (1885). Einleitung zu den Beiträgen zur Entwicklungsmechanik des Embryo. *Z. Biol.*, *21*, 411–428.

Roux, W. (1888). Beiträge zur Entwicklungsmechanik des Embryo. V. Über die künstliche Hervorbringung halber Embryonen durch Zerstörung einer der beiden ersten Furchungszellen, sowie über die Nachentwicklung (Postgeneration) der fehlendenKörperhälfte. *Ges. Abh.*, *2*, 419–521.

Roux, W. (1893). Über die Specification der Furchungszellen und über die bei der Postgeneration und Regeneration anzunehmenden Vorgänge. *Ges. Abh.*, *2*, 872–919.

Roux, W. (1897). *Programm und Forschungsmethoden der Entwicklungsmechanik der Organismen*. W. Engelmann, Leipzig.

Rudnick, D. (1945). Limb-forming potencies of the chick blastoderm, including notes on associated structures. *Trans. Conn. Acad. Arts Sci.*, *36*, 353–377.

Saxén, L. and Toivonen, S. (1961). The two-gradient hypothesis in primary induction. The combined effect of two types of inductors mixed in different ratios. *J. Embr. Exp. Morphol.*, *9*, 514–533.

Saxén, L. and Toivonen, S. (1962). *Primary Embryonic Induction*. Logos Press, London.

Schmidt, G. A. (1933). Schnürungs- und Durchschneidungsversuche am Anurenkeim. *Roux' Arch. f. Entw. mech.*, *129*, 1–44.

Spemann, H. (1898). Über die erste Entwicklung der Tuba Eustachii und des Kopfskeletts von *Rana temporaria*. *Zool. Jahrb., Anat. Ontog.*, *11*, 1–30.

Spemann, H. (1901a). Entwicklungsphysiologische Studien am Tritonei I. *Roux' Arch. f. Entw.mech.*, *12*, 224–264.

Spemann, H. (1901b). Über Correlationen in der Entwicklung des Auges. *Verhandl. Anat. Ges.*, *15*, 61–79.

Spemann, H. (1902). Entwicklungsphysiologische Studien am Tritonei II. *Roux' Arch. f. Entw. mech.*, *15*, 448–534.

Spemann, H. (1903a). Entwicklungsphysiologische Studien am Tritonei III. *Roux' Arch. f. Entw.mech.*, *16*, 551–631.

Spemann, H. (1903b). Über Linsenbildung bei defekter Augenblase. *Anat. Anz.*, *23*, 457–464.

Spemann, H. (1904). Über experimentell erzeugte Doppelbildungen mit zyklopischem Defekt. *Zool. Jahrb. Suppl. 7*, 429–470.

Spemann, H. (1906). Über eine neue Methode der embryonalen Transplantation. *Verh. d. deutschen Zool. Ges.*, 195–202.

Spemann, H. (1910). Die Entwicklung des invertierten Hörgrübchens zum Labyrinth. *Roux' Arch. f. Entw. mech,. 30*, 437–458.

Spemann, H. (1912a). Zur Entwicklung des Wirbeltierauges. *Zool. Jahrb., Abt. allg. Zool. Phys. d. Tiere, 32*, 1–98.

Spemann, H. (1912b). Über die Entwicklung umgedrehter Hirnteile bei Amphibienembryonen. *Zool. Jahrb. Suppl. 15*, 1–48.

Spemann, H. (1918). Über die Determination der ersten Organanlagen des Amphibienembryo, I–VI. *Roux' Arch. f. Entw.mech., 13*, 448–555

Spemann, H. (1919). Experimentelle Forschungen zum Determinations- und Individualitäts-problem. *Naturwissenschaften, 7*, 581–591.

Spemann, H. (1921). Die Erzeugung tierischer Chimären durch heteroplastische Transplantation zwischen *Triton cristatus* und *taeniatus*. *Roux' Arch. f. Entw.mech., 48*, 533–570.

Spemann, H. (1923). Zur Theorie der tierischen Entwicklung. *Rektoratsrede*, 1–16. Speyer und Kaerner, Freiburg im Breisgau.

Spemann, H. (1924a). Vererbung und Entwicklungsmechanik. *Naturwissenschaften, 12*, 65–79.

Spemann, H. (1924b). Über Organisatoren in der tierischen Entwicklung. *Naturwissenschaften, 12*, 1092–1094.

Spemann, H. (1927). Neue Arbeiten über Organisatoren in der tierischen Entwicklung. *Naturwissenschaften, 15*, 946–951.

Spemann, H. (1931a). Über den Anteil von Implantat und Wirtskeim an der Orientierung und Beschaffenheit der induzierten Embryonalanlage. *Roux' Arch. f. Entw. mech., 123*, 389–517.

Spemann, H. (1931b). Das Verhalten von Organisatoren nach Zerstörung ihrer Struktur. *Verh. d. deutschen Zool. Ges.*, 129–132.

Spemann, H. (1936). *Experimentelle Beiträge zu einer Theorie der Entwicklung.* Verlag Julius Springer, Berlin.

Spemann, H. (1938). *Embryonic Development and Induction.* Yale University Press, New Haven.

Spemann, H. (1941). Walther Vogt zum Gedächtnis. *Roux' Arch. f. Entw.mech., 141*, 1–14.

Spemann, H. (1942). Über das Verhalten embryonalen Gewebes im erwachsenen Organismus. *Roux' Arch f. Entw. mech., 141*, 693–769.

Spemann, H. (1943). *Forschung und Leben*, ed. F. W. Spemann. J. Engelhorn Nachf. Adolf Spemann, Stuttgart.

Spemann, H. und Bautzmann, E. (1927). Über Regulation von Triton-Keimen mit überschüssigem und fehlendem medianem Material. *Roux' Arch f. Entw. mech., 110*, 557–577.

Spemann, H. und Falkenberg, H. (1919). Über asymmetrische Entwicklung und *Situs inversus viscerum* bei Zwillingen und Doppelbildungen. *Roux' Arch. f. Entw. mech., 45*, 371–422.

Spemann, H., Fischer, E. G. und Wehmeier, E. (1933). Fortgesetzte Versuche zur Analyse der Induktionsmittel in der Embryonalentwicklung. *Naturwissenschaften, 21*, 505–506.

Spemann, H. und Geinitz, B. (1927). Über Weckung organisatorischer Fähigkeiten durch Verpflanzung in organisatorische Umgebung. *Roux' Arch. f. Entw. mech., 109*, 129–175.

Spemann, H. und Mangold, H. (1924). Über Induktion von Embryonalanlagen durch Implantation artfremder Organisatoren. *Roux' Arch. f. Entw. mech., 100*, 599–638.

Spemann, H. und Schotté, O. (1932). Über xenoplastische Transplantation als Mittel zur Analyse der embryonalen Induktion. *Naturwissenschaften, 20*, 463–467.

Spofford, W. R. (1948). Observations on the posterior part of the neural plate in *Amblystoma*. II. The inductive effect of the intact posterior part of the chorda-mesodermal axis on competent prospective ectoderm. *J. Exp. Zool., 107*, 123–164.

Takata, C. (1960). The differentiation *in vitro* of the isolated endoderm under the influence of the mesoderm in *Triturus pyrrhogaster. Embryologia, 5*, 38–70.

Ten Cate, G. (1956). The intrinsic embryonic development. *Verh. Konink. Nederl. Akad. Wetensch. Nat., 51*, 3–257.

Tiedemann, H. (1963). The role of regional specific inducers in the primary determination and differentiation of amphibia. In: *Biological Organization at the Cellular and Supercellular Level,* ed. R. J. C. Harris, pp. 183–209. Academic Press, New York.

Tiedemann, H. (1967). Biochemical aspects of primary induction and determination. In: *The Biochemistry of Animal Development,* Vol. 2, ed. R. Weber, pp. 3–55. Academic Press, New York and London.

Tiedemann, H. (1975). Substances with morphogenetic activity in differentiation of vertebrates. In: *The Biochemistry of Animal Development,* Vol. 3, ed. R. Weber, pp. 257–292. Academic Press, New York and London.

Tiedemann, H. (1978). Chemical approach to the inducing agents. In: *Organizer—a Milestone of Half-Century from Spemann,* ed. O. Nakamura and S. Toivonen, pp. 91–117. Elsevier/ North-Holland Biomedical Press, Amsterdam.

Tiedemann, H. und Tiedemann, H. (1959). Versuche zur Gewinnung eines mesodermalen Induktionsstoffes aus Hühnerembryonen. *Hoppe-Seylers Zeitschr. f. Physiol. Chemie, 314,* 156–176.

Tiedemann, H. und Tiedemann, H. (1964). Das Induktionsvermögen gereinigter Induktionsfaktoren im Kombinationsversuch. *Rev. Suisse de Zool., 71,* 117–137.

Toivonen, S. (1938). Spezifische Induktionsleistungen von abnormen Induktoren im Implantatversuch. *Ann. Zool. Soc. Zool.-Bot. Fenn. Vanamo, 6,* 1–12.

Toivonen, S. (1940). Über die Leistungspezifität der abnormen Induktoren im Implantatversuch bei *Triton. Ann. Acad. Sci. Fenn. Ser. A, 55,* 1–150.

Toivonen, S. (1949). Zur Frage der Leistungsspezifität abnormer Induktoren. *Experientia, 5,* 323–325.

Toivonen, S. (1953a). Bone-marrow of the guinea pig as a mesodermal inductor in implantation experiments with embryos of *Triturus. J. Embr. Exp. Morphol., 1,* 97–104.

Toivonen, S. (1953b). Knochenmark als mesodermaler Induktor im Implantatversuch bei *Triturus. Arch. Soc. Zool.-Bot. Fenn. Vanamo, 7,* 113–121.

Toivonen, S. (1954). The inducing action of the bone-marrow of the guinea pig after alcohol and heat treatment in implantation and explantation experiments with embryos of *Triturus. J. Embr. Exp. Morphol., 2,* 239–244.

Toivonen, S. (1961). An experimentally produced change in the sequence of neuralizing and mesodermalizing inductive actions. *Experientia, 17,* 87–88.

Toivonen, S. (1978). Regionalization of the embryo. In: *Organizer—a Milestone of Half-Century from Spemann,* ed. O. Nakamura and S. Toivonen, pp. 119–156. Elsevier/North-Holland Biomedical Press, Amsterdam.

Toivonen, S. and Saxén, L. (1955a). The simultaneous inducing action of liver and bone-marrow of the guinea pig in implantation and explantation experiments with embryos of *Triturus. Exp. Cell. Res. Suppl. 3,* 346–357.

Toivonen, S. und Saxén, L. (1955b). Über die Induktion des Neuralrohrs bei Trituruskeimen als simultane Leistung des Leber- und Knochenmarkgewebes vom Meerschweinchen. *Ann. Acad. Sci. Fenn. Ser. A. Biol.,* 1–29.

Toivonen, S. and Saxén, L. (1968). Morphogenetic interaction of presumptive neural and mesodermal cells mixed in different ratios. *Science, 159,* 539–540.

Twitty, V. (1955). Eye. In: *Analysis of Development,* ed. B. H. Willier, P. Weiss, and V. Hamburger, pp. 402–414. W. B. Saunders Company, Philadelphia and London.

Vogt, W. (1922). Operativ bewirkte Exogastrulation bei *Triton* und ihre Bedeutung für die Theorie der Wirbeltiergastrulation. *Anat. Anz. 55, Erg. Heft.,* 53–64.

Vogt, W. (1923). Morphologische und physiologische Fragen der Primitiventwicklung. Versuche zu ihrer Lösung mittels vitaler Farbmarkierung. *Ber. Ges. Morph., Physiol., München, 35,* 22–32.

Vogt, W. (1927). Über Hemmung der Formbildung an einer Hälfte des Keimes (Nach Versuchen an Urodelen). *Anat. Anz. 63, Erg. Heft.,* 126–139.

Vogt, W. (1928). Mosaikcharakter und Regulation in der Frühentwicklung des Amphibieneies. *Verh. d. deutschen Zool. Ges.,* 26–70.

Vogt, W. (1929a). Hans Spemann zum 60. Geburstag. *Roux' Arch. f. Entw. mech., 116,* XIII–XV.

Vogt, W. (1929b). Gestaltungsanalyse am Amphibienkeim mit örtlicher Vitalfärbung. II. Gastrulation und Mesodermbildung bei Urodelen und Anuren. *Roux' Arch. f. Entw. mech., 120,* 384–706.

Waddington, C. H. (1933). Induction by the primitive streak and its derivatives in the chick. *J. Exp. Biol., 10,* 38–46.

Waddington, C. H., Needham, J. und Needham, D. M. (1933). Beobachtungen über die physikalisch-chemische Natur des Organisators. *Naturwissenschaften, 21,* 771–772.

Watterson, R. L. (1955). Selected invertebrates. In: *Analysis of Development,* ed. B. H. Willier, P. Weiss, and V. Hamburger, pp. 315–336. W. B. Saunders Company, Philadelphia and London.

Wehmeier, E. (1934). Versuche zur Analyse der Induktionsmittel bei der Medullarplatteninduktion von Urodelen. *Roux' Arch. f. Entw. mech., 132,* 384–423.

Weismann, A. (1892). *Das Keimplasma. Eine Theorie der Vererbung.* Gustav Fischer, Jena (English ed. 1893).

Weiss, P. (1925). Unabhängigkeit der Extremitätenregeneration vom Skelett (bei *Triton cristatus*). *Roux' Arch. f. Entw. mech., 104,* 359–394.

Weiss, P. (1939). *Principles of Development.* Henry Holt and Company, New York.

Willier, B. H. and Oppenheimer, J. (1974). *Foundations of Experimental Embryology,* Second Edition. Hafner Press, New York.

Winkler, H. (1907). Über Propfbastarde und pflanzliche Chimaeren. *Ber. deut. Bot. Ges., 25,* 568–576.

von Woellwarth, C. (1952). Die Induktionsstufen des Gehirns. *Roux' Arch. f. Entw. mech., 145,* 582–668.

Woerdeman, M. W. (1933a). Über den Glykogenstoffwechsel des Organisationszentrums in der Amphibiengastrula. *Kon. Akad. Wetensch. Amsterdam, Proc., 36,* 189–193.

Woerdeman, M. W. (1933b). Über den Glykogenstoffwechsel tierischer Organisatoren. *Kon. Akad. Wetensch. Amsterdam, Proc., 36,* 423–426.

Woerdeman, M. W. (1933c). Ueber die chemischen Prozesse bei der embryonalen Induktion. *Kon. Acad. Wetensch. Amsterdam, Proc. 36,* 842–849.

Wolpert, L. (1970). Positional information and pattern formation. In: *Towards a Theoretical Biology,* ed. C. H. Waddington, pp. 198–230. Aldine Publishing Company, Chicago.

Wolpert, L. (1971). Positional information and pattern formation. *Curr. Topics Dev. Biol., 6,* 183–224.

Yamada, T. (1950). Regional differentation of the isolated ectoderm of the *Triturus* gastrula induced through a protein extract. *Embryologia, 1,* 1–20.

Yamada, T. (1959). A progressive change in regional inductive effects of the bone-marrow caused by heat-treatment. *Embryologia, 4,* 175–190.

Yamada, T. (1961). A chemical approach to the problem of the organizer. *Adv. Morphogen., 1,* 1–53.

Yntema, C. L. (1950). An analysis of induction of the ear from foreign ectoderm in the salamander embryo. *J. Exp. Zool., 113,* 211–244.

Yntema, C. L. (1955). Ear and nose. In: *Analysis of Development,* ed. B. H. Willier, P. Weiss, and V. Hamburger, pp. 415–428. W. B. Saunders Company, Philadelphia and London.

Index